OCCUPATIONAL SAFETY AND HEALTH

A Series of Reference Books and Textbooks
on Occupational Hazards ● Safety ● Health ●
Fire Protection ● Security ● and Industrial Hygiene

Series Editor
ALAN L. KLING
Loss Prevention Consultant
Jamesburg, New Jersey

1. Occupational Safety, Health, and Fire Index   *David E. Miller*
2. Crime Prevention Through Physical Security   *Walter M. Strobl*
3. Fire Loss Control   *Robert G. Planer*
4. MORT Safety Assurance Systems   *William G. Johnson*
5. Management of Hotel and Motel Security   *Harvey Burstein*
6. The Loss Rate Concept in Safety Engineering   *R. L. Browning*
7. Clinical Medicine for the Occupational Physician   *edited by Michael H. Alderman and Marshall J. Hanley*
8. Development and Control of Dust Explosions   *John Nagy and Harry C. Verakis*
9. Reducing the Carcinogenic Risks in Industry   *edited by Paul F. Deisler, Jr.*
10. Computer Systems for Occupational Safety and Health Management   *Charles W. Ross*
11. Practical Laser Safety   *D. C. Winburn*
12. Inhalation Toxicology: Research Methods, Applications, and Evaluation   *edited by Harry Salem*
13. Investigating Accidents with STEP   *Kingsley Hendrick and Ludwig Benner, Jr.*
14. Occupational Hearing Loss   *Robert Thayer Sataloff and Joseph Sataloff*
15. Practical Electrical Safety   *D. C. Winburn*

Other Volumes in Preparation

Practical
Electrical
Safety

To

Cecile, Evelyn, Laurel
Ellabelle, and Delores,
my wonderful sisters

Preface

Electrocution, death by electrical shock, is one of the leading causes of industrial fatalities, according to the National Safety Council. Injuries caused by electrical accidents in the home and workplace also rate high in serious biolgoical damage and lost-time statistics. Many users of electrical appliances, equipment, and systems either falsely assume adequate knowledge of the hazards involved, or they use such equipment without being properly trained in its safe use.

As an engineer involved in laboratory research using many varieties of electrical and electronic equipment and apparatus, I have experienced several incidents that were unexpected. Because electrical safety was incidental to the function of the apparatus, I was not offered detailed indoctrination in the potential hazards. It is the purpose of this book to provide the user with specific information in one reference. Details about the safe use of electrical equipment may be available elsewhere, but until now there has been no single reference designed to cover the potential hazards in power distribution, electrical construction (including salient portions of the National Electrical Code), and the use of electrical equipment (including relevant Occupational Safety and Health Administration requirements).

I hope that this book will serve as a text in industrial safety curricula, as a guide for industrial safety engineers to control industrial electrical hazards, and as a reference in electrical safety training.

Much of the information contained in this work was developed during my thirty-five years as a supervisor and safety officer in materials and laser research and technology at the Los Alamos National Laboratory.

D. C. Winburn

Contents

Preface	v
1. Introduction	1
References	3
2. Effects of Electrical Current in the Human Body	5
Perception of Electrical Shocks	5
Let-Go Current Determinations	7
Effects of High Currents	10
Burns from Electric Shock	11
Rescue and Resuscitation of Electric Shock Victims	12
Summary	14
References	16
3. Controlling Electrical Hazards	17
Workplace Hazards	22
Controlling Electrical Hazards in Construction	36
Controlling Electrical Hazards in the Home	42

4.	National Electrical Code 1987	47
	Scope of the Code	48
	Requirements for Electrical Installations	50
	Wiring Design and Protection	54
	Wiring Methods and Materials	54
	Equipment for General Use	55
	Special Occupancies	57
	Special Equipment	58
	Special Conditions	60
	Communications Systems	61
	Tables and Examples	63
5.	National Electric Safety Code	65
	Purpose and Scope of NESC	67
	Installation and Maintenance of Electric Supply Stations and Equipment (Part 1)	78
	Installation and Maintenance of Overhead Electric Supply and Communication Lines (Part 2)	82
	Installation and Maintenance of Underground Electric-Supply and Communication Lines (Part 3)	89
	Operation of Electric-Supply and Communications Lines and Equipment (Part 4)	92
6.	Federal Regulations	127
	General	129
	Design Safety Standards for Electrical System	129
	Definitions	178
7.	Case Histories and Electrical Workers Survey	199
	Case Histories	199
	Electrical Workers Survey	201
8.	Instituting an Effective Electrical Safety Management Program in Research and Development Facilities	223
	Scope and Applications (Part I)	224
	General Criteria for Electrical Safety Administration (Part II)	225
	General Safety Criteria for Design and Construction (Part III)	228

CONTENTS

General Operating Safety Criteria (Part IV)	231
Safety Criteria for Working on Electrical Equipment (Part V)	232
Electrical Hazards and Emergency Procedures (Part VI)	236
Equipment Safety Exhibits (Part VII)	239
Additional Chemical, Fire, and Other Hazards Associated with Operation of Electrical R&D Equipment (Part VIII)	261
Appendix A Definition of Terms	266
Appendix B Bibliography	268
Appendix C Electrical Safety Criteria Committe Membership (Part IX)	270
Index	*273*

Practical
Electrical
Safety

1
Introduction

Though he played a considerable role in harnessing electricity to be used to man's advantage, Benjamin Franklin became so concerned about this biological hazard that he conducted experiments with chickens, using their heads as conductors. The chickens either died from the shock or showed essentially no aftereffects of the experiment.

These early experiments were important in establishing the lethal effects of electrical shock. According to the *Encyclopaedia Britannica* (1), three human physiological conditions can result from electrical shock, depending on the circuit's route through the body and the amount of current transmitted. *Muscular contraction* results from a through-body circuit, and, if the lungs are involved, voluntary respiration can be halted and asphyxiation takes place. If the heart is involved in the circuit, *fibrillation*, or irregular heart beats, can occur, and ultimately heart failure can result. Conversely, irregular heart beat can often be treated with electrical shock to restore rhythmic, regular heart palpitations.

The definition of *electrocution*, according to Wester (2), is simply death by electrical shock. Injury produced by other industrial hazards, such as mechanical or chemical accidents, can be fatal, also but, if not fatal, then the resulting physical condition of the injured party can vary considerably. Apparently not so for electrical shock; the result is either death or no lasting effect, except maybe for the contact surfaces which may develop blisters or sores that generally heal with time. Therefore, the key to prevention of electrical shock is to emphasize the inaccessability of the circuit.

Franklin's concern with electrical safety accompanied his pioneering work with an international collection of scientists, including Alessandro Volta, of Italy, and Michael Faraday of Britain. It was Faraday who first demonstrated the relationship between electricity and magnetism in 1831, and his experiments provided the point of departure for both the mechanical generation of electric current, previously available only from chemical reactions within voltaic piles, or batteries, and the utilization of such currents in electric motors. Both generators and motors underwent substantial development in the middle decades of the nineteenth century. Britain had a well established tradition of steam power, coal, and coal gas, so development of a market for electricity fell on the continental Europe and North American experimenters. In the United States, it was Thomas Edison who applied his inventive genius to finding fresh uses of electricity. Filament lamps were developed, but lighting alone could not provide a feasible economic impact for electricity, because its use was confined to the hours of darkness. The development of electric traction made urban electric tramways popular and coincided with the widespread construction of power-generating equipment in the late 1880s. The subsequent spread of this form of energy is one of the most remarkable technological success stories of the twentieth century, but most of the basic techniques of generation, distribution, and utilization had been mastered by the end of the nineteenth century.

In 1881, the National Association of Fire Engineers met in Richmond, Virginia. From this meeting came a proposal that served as a basis for the first National Electric Codes (3), covering such items as the use of the insulated conduit, the use of single disconnect devices, and identification of the white wire. In 1911, the National Fire Protection Association (NFPA) assumed sponsorship and control of the National Electrical Code. Since 1920, the National Electrical Code has also received official endorsment by the American National Standards Institute, and the NFPA has maintained its capacity as Administrative Sponsor. Since that date, the Committee has been identified as "ANSI Standards Committee C1." The 1984 Edition of the Code (NFPA 70-1984) was adopted by the Fire Protection Association and approved by the American National Standards Institute as

INTRODUCTION

ANSI/NFA 70-1984. This Code is advisory as concerns the NFPA and ANSI, but is useful in legal and regulatory matters and in the interest of life and property protection. Portions of the Code that relate to safety will be included in this work as well as excerpts of the Code of Federal Regulations that interprets and applies the Williams-Steiger Occupational Safety and Health Act of 1970, commonly referred to as the OSHA Act. Also discussed is the American National Standard ANSI C2-1987, the National Safety Code (4), serviced by the IEEE, Inc. as Secretariat.

REFERENCES

1. *Encyclopaedia Britannica*, 15th ed., Helen Hemingway Benton, Publisher, 425 N. Michigan Avenue, Chicago, IL 60611, 1973-74.
2. *Webster's Third New International Dictionary.* © G. & C. Merriam Co. Published by Encyclopaedia Britannica, Inc. Chicago, 1976.
3. National Electric Code. National Fire Protection Association, Quincy, MA, 1984 edition.
4. American National Standard ANSI C2-1987, National Electrical Safety Code. The Institute of Electrical and Electronics Engineers, Inc. Distributed in cooperation with Wiley-Interscience, a division of John Wiley and Sons, Inc., New York City.

2
Effects of Electrical Current in the Human Body

Electricity has become an essential part of modern life, both at home and in the workplace. The small number of electrical accidents that prove fatal, worldwide, about 1000 per year over the past 15 years (1), is due to the general recognition that electricity is inherently dangerous. However, users must be alert to potentially dangerous conditions that could lead to electrical accidents involving the possible passage of electrical current through human tissue.

Knowledge of the possible effects of electrical current on man is the starting point at which to incorporate safety into the design of electrical equipment. This knowledge is essential in explaining and justifying safety regulations, safe work conditions, and in rescue and resuscitation methods for electric shock victims. This information is vital in educating the general public in the proper use of appliances and machines. Such knowledge is also valuable to the physician in diagnosing and treating victims of electrical-related accidents.

PERCEPTION OF ELECTRICAL SHOCK

Although high voltage often produces spectacular and awesome destruction of tissue at contact locations, it is generally believed that the detrimental effects of electric shock are due to the current actually flowing through the body. Even though Ohm's law (I = E/R) applies, it is often difficult to correlate voltage with damage to the body because of the large variations in contact resistance usually present in accidents. For this reasons, criteria of the danger from electric shock are based solely upon current. The electric shock hazards of both high-voltage and low-voltage circuits are derived from known effects of electric currents as determined from low-voltage experiments. Experiments involving currents large enough to cause sudden death cannot be made on human beings, and resort is made to animal experimentation. Although translation of results obtained in animals to humans is conjectural, in many instances analysis of human accidents has permitted satisfactory correlation to be made.

When electricity passes through some portion of the human body, it is important to distinguish between the direct effects of electric current and general shock to the nervous system, which is commonly called "shock." Because of the wide variation in the physical condition of individuals, it is impossible to justify any electric shock as safe for *all* individuals. Frequent accounts of fatalities ascribed to heart failure due to overexcitement, intense emotion, fear, or shock appear in the press. For such susceptible persons, it is possible that contact with any electric circuit which permits currents in excess of the threshold of sensation might result in fatality. This possibility must be recognized, and an occasional death is to be expected from casual contact involving electric currents known to be safe for most normal individuals. In such cases, death must be considered to be due to shock to the nervous system and not to the primary effects of electric current. In contrast, establishment of reasonably safe currents is a vital problem of practical importance for the vast majority of normal individuals.

According to one authority (2), the ratio of fatalities to injuries for electric shock accidents is very high in comparison to the corresponding figure for all other accidents. However, it is indeed fortunate that victims who survive the first effects of serious electrical accidents often recover without permanent disability. (This condition was observed by Benjamin Franklin with his live chickens!)

At first, tingling sensations result from barely perceptible currents, proceed with reactions produced by currents of increasing intensity, and ultimately result in various causes of electrocution. Effects caused by high-frequency currents include burns and blisters.

EFFECTS OF ELECTRICAL CURRENT ON HUMANS

The highly developed human nervous system makes man extremely sensitive to even very small electric currents. An almost unlimited number of thresholds of sensation could be defined depending upon the locations selected for applying contacts to the body. The effects of electric stimulation depend, to a very considerable extent, upon the type of contacts, whether the contacts are firm and involve an appreciable area, or whether they are point contacts. Currents almost too small to measure produce severe piercing pain when they flow in an open cut or wound.

Perception of current on the hand is important, as it is essential that the user not feel the sensation of shock when using home appliances, hand tools, or surgical instruments. Shocks due to currents only slightly in excess of the threshold of perception are usually considered annoying rather than dangerous. However, they are startling when they are not anticipated. Such shocks may cause an involuntary movement or loss of balance, and the ensuing fall or movement might result in painful injury. Obviously, the slightest shock to a surgeon during an operation might be disastrous to a patient.

LET-GO CURRENT DETERMINATIONS

With increasing alternating current, the sensations of tingling give way to contractions of the muscles. The muscular contractions and accompanying sensations of heat increase as the current is increased. Sensations of pain develop, and voluntary control of the muscles that lie in the current pathway becomes increasingly difficult. Finally a value of current is reached for which the subject cannot release his grasp of the conductor. At this point, he is said to "freeze" to the circuit. The maximum current a person can tolerate when holding a conductor in one hand and still let go of the conductor by using the muscles directly stimulated by that current is called his "let-go" current. Let-go currents are important, as experience has shown that an individual can withstand, with no ill after effects, except possibly sore muscles, repeated exposure to his let-go current for at least the time required for him to release the conductor.

Sixty-cycle let-go currents were measured on women and men in experiments by Dalziel (2). Subjects ranged in age from the late teens to the early twenties. The women were light in stature and obviously not accustomed to hard physical work, and their forearm muscles were not particularly well developed. The results are probably representative for the sedentary type. However, from observation of the reactions of the subjects having the greatest muscular development, it is possible that values were considerably lower than those which would have been obtained had a group of mature, healthy women accustomed to physical labor been used. Results based on these data, therefore, should be conservative and on the side of safety.

The average value, or let-go threshold, was established at 15.87 and 10.5 mA for males and females, respectively. The ratio of the let-go thresholds, females to males, is approximately 2/3. This ratio is frequently used in estimating let-go currents for women for other frequencies and waveforms. It is believed that the lower value for women is due to their somewhat poorer muscular development rather than to any difference due to sex. No satisfactory theory has been advanced to explain why the female to male perception and let-go threshold ratios are seemingly identical.

Tests using gradually increasing *direct current* (dc) produced sensations of internal heating rather than severe muscular contractions. Sudden changes in the current magnitude produced powerful muscular contractions, and interruption of the current produced a very severe shock. Muscular reactions when the test electrode was released at the higher values were objectionable and sooner or later all subjects declined to attempt higher currents. Tests were made on 28 men; in each case little difficulty was experienced in releasing the electrode. The maximum current a subject would withstand before releasing the electrode was termed his "release" current, since this represents a psychological limit rather than the physiological limit of the let-go tests. Because of the relatively small number of subjects used, the average dc release current for a large number of men was computed as equal to the average dc release current times the ratio of the average 60-cycle let-go current for 134 men to the average 60-cycle let-go current of the 28 man sample. Thus, the probable average dc release current for a large group of men is estimated at 76.1 mA. Assuming a similar distribution for women, the corresponding dc release current would be 50.7 mA.

At the conclusion of the 60-cycle let-go current tests on the women, the experiments were terminated with one or two release tests using direct current. After one or two preliminary trials, the current was increased to a maximum of 35 mA, which each subject released without complaint or difficulty. In one case, a woman was tested at the same time as the dc release tests were being made on the men. She released 56 mA direct current before refusing more.

Although the deleterious effects of electric shock are due to the current actually flowing through the human body, in accidents the voltage of the circuit is usually the only electrical quantity known with certainty. While current and voltage are related by Ohm's law, the great variances in skin and contact resistances are so unpredictable that let-go voltages are relatively meaningless. On very high voltage circuits, skin and contact resistances break down instantly, and thus they play only a minor role in limiting the current received by a victim. However, on the lower voltages the resistances at contact locations become of increasing importance, and these resistances are of paramount importance on very low voltage circuits. Obviously, wet contacts create a most dangerous condition for receiving an electric shock, and let-go voltages under these conditions may be of limited interest.

From these and other similar tests (3) it is concluded that reasonably safe 60-cycle let-go voltages hand-to-hand are about 21 volts, and hand-to-both-feet, ankle deep in salt water, 10 volts. Reasonably safe dc release voltages are 104 volts hand-to-hand, and 51 volts hand-to-both-feet ankle deep in salt water (3).

According to Dalziel (2), man's greater tolerance to electric current with increasing frequency suggests that the effect of higher harmonics, or repetitive pulses superimposed on low-frequency alternating current, would be less and less as the frequency of the superimposed current was increased. At very high frequency, it might be expected that the sensation of heat would be largely masked out by the pain and the muscular stimulation of the lower frequency components.

The higher 60-cycle let-go currents in the tests mentioned were frequently sufficient to stop breathing during the period the current was flowing across the chest, and the reactions at the instant of current interruption during the dc release tests occasionally threw the subject a considerable distance. Muscular reactions during accidents frequently cause fractures, and contractions resulting when a victim grasps bare overhead wires may be sufficient to freeze him suspended to the circuit in spite of his struggle to drop free. In many accidents, a victim frees himself by breaking the conductor, or his body weight may assist him in interrupting the circuit; however, fortuitous circumstances must not be relied upon to provide safety to human life. Currents only slightly in excess of one's let-go current value are very painful, frightening, and hard to endure for even a short time. Failure to interrupt the current promptly is accompanied by a rapid decrease in muscular strength due to the pain and fatigue associated with the accompanying severe involuntary muscular contractions, and it would be expected that the let-go ability would decrease rapidly with the duration of contact. Prolonged exposure to currents only slightly in excess of a person's let-go limit may produce exhaustion, asphyxia, collapse, and unconsciousness followed by death.

EFFECTS OF HIGH CURRENTS

Currents considerably in excess of those required to cause a stoppage of breathing due to excessive contraction of the chest muscles may produce temporary paralysis of respiration due to action on the nerves. It has been known for some time that respiration might be inhibited by currents passing through the respiratory nerve center located in the base of the brain. These victims are almost always unconscious and appear to be dead. The paralysis may last for a considerable time after interruption of the current and resuscitation must be applied *immediately* to prevent asphyxial death. Often the paralysis disappears in a few minutes or hours, and continued application of artificial respiration saves the victim. Mere cessation of natural breathing is not likely to produce serious aftereffects or permanent damage if artificial respiration is applied immediately, as evidenced by the many persons who have been resuscitated successfully. Unfortunately, little is known regarding the magnitude of the currents required to produce respiratory inhibition or unconsciousness.

With contacts on external parts of the body and with the current pathway involving the chest, currents considerably in excess of those sufficient to just stop breathing due to muscular contraction of the chest muscles may produce respiratory inhibition, heart block, ventricular fibrillation, or irreversible damage to vital parts of the nervous system. Ventricular fibrillation is nearly always fatal and is commonly called instantaneous electrocution. Respiratory inhibition, heart block, or serious damage to the nervous system requires shocks of considerably greater intensity than those required to produce ventricular fibrillation. For this reason, the fibrillating threshold is of extreme importance, as no person should knowingly be subjected to shocks of this magnitude. Experimental work on human hearts is obviously impossible, and resort must be made to animal experimentation. Evaluation of the fibrillating threshold is difficult not only because of the uncertainty of relating the results obtained on animals to humans, and because the susceptibility of the heart to fibrillate varies at different parts of its cycle, but also because of the limited amount of data available.

For short shocks, the susceptibility of the heart to fibrillate increases with increasing current until a most dangerous current is reached—then decreases. At relatively high currents, the likelihood of producing ventricular fibrillation is almost negligible. The explanation of the phenomenon is that very high currents paralyze the nerve centers in the heart, the heart is contracted and silenced, and fibrillation is prevented. Death is inevitable if the shock is of appreciable duration; however, if the shock is of short duration, and if the heart has not been damaged, interruption of the current may be followed by a spontaneous resumption of its normal rhythmic contractions. This is the explanation offered for many accident cases in which victims apparently withstood relatively high currents. This phenomenon is the basis for countershock, or defibrillation shock treatment, to arrest ventricular fibrillation. It is also believed that the message and accompanying stimulation of the heart produced by application of approved methods of manual respiration may be beneficial in assisting the heart to regain its normal rhythm.

BURNS FROM ELECTRIC SHOCK

Another important effect of electric shock is the electric burn or blister; in fact, burns and blisters are the most common result of electrical accidents. Burns are of two types, thermal and electric. Electric burns are produced by the current flowing through the flesh of the tissues, especially at point contacts where high current densities are present. Electric burns are often caused when high-voltage arcs or sparks bridge the gap between the energized conductor and the body (although electric burns may be produced at poor or varying contacts during low-voltage accidents). Blisters or burns destroy the protective resistance of the epidermis, thereby permitting greater currents to flow. Should the current flow for a long enough time, the heat developed by the current may raise the temperature of the body to a value sufficient to cause death.

Electric burns are slow to heal but seldom become infected. Thermal burns are the result of high temperatures in close proximity to the body, such as produced by an electric arc, vaporized metals, or hot gases released by the arc, or by overheated conductors caused by short circuits. These burns are similar to burns and blisters produced by any high temperature source. Currents of the let-go level are more than sufficient to produce deep burns, and both types of burns may be produced simultaneously. Any serious burn should receive prompt medical attention.

The reactions produced by high-frequency is increased, and sensations of heat become predominant. At low values of current, apparently subjective sensations of heat are caused by stimulation of the heat-sensitive mechanisms, but at higher currents the sensations are undoubtedly due to the actual increase in body temperature. As the frequency increases, the tingling sensations produced by currents of the perception level change to sensations of heat. The transition occurred between 100 and 200 kc for the 25 subjects tested (2). Currents barely perceptible above the transition manifest themselves by sensations of heat only. The reactions associated with let-go currents remain essentially the same up to 5 kc. At 10 kc, sensations of heat are definite, the time required to release the conductor increases appreciably, and the muscular reactions are quite sluggish. Many authorities are of the opinion that heat or burns are the only effects of high-frequency currents; however, the high-frequency domain is largely unexplored in this respect, and there is much to be learned regarding the electric effects produced by very high frequency and extremely short-pulsed current.

RESCUE AND RESUSCITATION OF ELECTRIC SHOCK VICTIMS

No discussion of electric shock would be complete without mention of rescue and resuscitation for victims of serious electric shock accidents. *Rescue* the victim from the circuit *promptly* and *safely*. In many cases, the victim may remain in contact with the circuit because of his inability to let go of the conductor, or due to unconsiousness. *Apply artificial respiration* if the victim is not breathing or if he appears not to be breathing. *Continue resuscitation* without interruption until the victim revives, until rigor mortis sets in, or until he or she is pronounced dead by a physician.

Approved methods of resuscitation have received universal acceptance because of successful field experience. The degree of success that can be attained in reviving victims is dependent upon two factors: (a) promptness in starting resuscitation efforts, and (b) maintaining an unobstructed air passage to the lungs. The mouth-to-mouth breathing method is the most effective and is described here in detail.

How to Restore the Breath of Life

After using first aid methods to determine whether the victim's heart is beating, check neck artery for pulse or whether victim is breathing (listen with ear over nose and mouth), follow these instructions closely.

1. Turn victim on back and quickly wipe out mouth. Place one hand under victim's neck and lift, tilting head back as far as possible with other hand. This provides air passageway.

2. If victim is not breathing, pinch nostrils shut. Place your mouth over victim's tightly and blow hard enough to make chest rise. (For very young children, cover both nose and mouth tightly with your mouth.)

3. Remove your mouth, allowing victim to exhale. Repeat blowing and removing your mouth, listening for victim's exhaling air. If you don't hear it, recheck head position. After four breaths, check for neck pulse by releasing hand clamping nose and using index fingers to feel for carotid pulse in neck artery under angle of lower jaw. If pulse is present, continue rescue breathing at 12 times a minute (every 5 seconds). For small child or infant blow gently 20 times a minute.

4. If no pulse is obtained, start dual procedures of breathing and cardiac compression. With victim's back on firm surface, place heel of one hand on center of lower breastbone with fingers away from chest and other hand on top. Gently rock forward, exerting considerable pressure down, to force blood out of the heart. Release pressure. Immediately return to breathing technique. Alternate breathing and chest massage every 5 seconds each. Check occasionally for pulse.

Caution: Small children and infants need much less pressure than adults on breastbone. Use only the heel of one hand and tips of index fingers. Exert only enough pressure to depress chest $1/2$ to $3/4$ of an inch. Give 80 to 100 compressions a minute with two breaths after each five.

(Courtesy of Metropolitan Life Insurance Company.)

> *Note*: Local Red Cross chapters and Heart Associations offer, without cost, complete courses in rescue and cardiopulmonary resuscitation (CPR). All citizens should attend first aid and CPR training classes, especially those individuals using extensive electrical appliances, equipment, or apparatus.

SUMMARY

A summary of the possible quantitative effects of electric currents on humans is given in Table 1. The data on which these results are based were collected over a period of several years and, as a result, the number of subjects used to establish a given threshold varied considerably. As mentioned previously, the perception threshold for women was assumed equal to 66 2/3% of that for men. The let-go threshold for women was determined for 60-cycle alternating current only; values for the other frequencies were taken at 66 2/3% of the corresponding values for men.

Obviously, no experimentation can be performed on humans to determine the current likely to produce instantaneous death. The values at the bottom of the table were derived from tests made on animals and are what might be termed "best estimates." They are to be regarded as possible values only and are subject to revision should more reliable data become available at some future date.

The known lethal effects of electric currents are summarized as follows:

1. If continued at length, currents in excess of an individual's let-go current may produce collapse, unconsciousness, and death. Although the causes of death are not known with certainty, asphyxiation or heart failure due to exhaustion and shock to the nervous system are prime suspects.
2. Currents flowing through the chest, head, or nerve centers controlling respiration may produce respiratory inhibition. This is caused by a nerve block which blocks the nerve impulses between the respiratory center and respiratory muscles. Respiratory inhibition is dangerous because paralysis of the respiratory organs may last for a considerable period even after interruption of the current. An approved method of artificial resuscitation must be applied *promptly* to prevent suffocation.

EFFECTS OF ELECTRICAL CURRENT ON HUMANS

Table 2.1 Quantitative Effects of Electric Current on Humans

	Milliamperes (thousandths of an ampere), mA					
	Direct current		Alternating current RMS values			
			60 Cycle		10,000 Cycles	
Effect	Men	Women	Men	Women	Men	Women
No sensation on hand	1	0.6	0.4	0.3	7	5
Slight tingling. Perception threshold	5.2	3.5	1.1	0.7	12	8
Shock — not painful and muscular control not lost	9	6	1.8	1.2	17	11
Painful shock — painful but muscular control not lost	62	41	9	6	55	37
Painful shock — let-go threshold	76	51	16.0	10.5	75	50
Painful and severe shock — muscular contractions, breathing difficult	90	60	23	15	94	63
Possible ventricular fibrillation from short shocks:						
Shock duration 0.03 sec	1300	1300	1000	1000	1100	1100
Shock duration 3.0 sec	500	500	100	100	500	500
Ventricular fibrillation — certain death	Multiply values immediately above by 2¾. To be lethal, short shocks must occur during susceptible phase of heart cycle					
Possible ventricular fibrillation from impulse shocks:						
DC short shocks and surge discharges	27.0 W-seconds					
Power-frequency short shocks and low-frequency oscillatory discharges	13.5 W-seconds					

3. Ventricular fibrillation is caused by moderately small currents which produce overstimulation rather than damage to the heart. When fibrillation occurs, the ventricles go into

asynchronous, or fibrillary, twitchings in contrast to their normal synchronous contractions, the rhythmic pumping action of the heart ceases, and death usually follows within minutes. The hearts of certain small animals recover from ventricular fibrillation spontaneously. However, the hearts of larger animals do not. It is believed that once ventricular fibrillation occurs in humans, it is unlikely to stop naturally before death. Prompt application of resuscitation insures the most favorable conditions in the event fibrillation should cease before death.

4. Heart block or suspension of heart action may be caused by relatively large currents. In cases where the shock is of short duration and where damage to the heart has not occurred, the heart may regain its normal rhythm automatically. It is believed that resumption of normal action is aided by the massaging action produced by approved methods of manual resuscitation.

5. Destruction of tissues due to high temperatures may cause complications leading to death.

6. Patients who have been revived sometimes die suddenly without apparent cause (delayed death). This may occur minutes, hours, or even days after the accident. This is thought to be due to (a) aggravation of pre-existing conditions, (b) the result of hemorrhages affecting vital centers, or (c) the effects of shock. Delayed death may also be due to burns or other complications.

7. Combinations of the above effects may occur simultaneously, or they may develop progressively, thereby making accurate diagnosis difficult.

REFERENCES

1. *Accident Facts*. National Safety Council, 425 N. Michigan Ave., Chicago, IL 60611.
2. Dalziel, C. F. The effects of electrical shock on man, *IRE Transactions on Medical Electronics*, (PGME-5), May 1956.
3. Dalziel, C. F., and Massoglia, F. P. Let-go currents and voltages, *Electric Arc and Resistance Welding-IV*, AIEE Special Publication, July 1954.

3
Controlling Electrical Hazards

Some employees work with electricity in a direct manner. Such is the case with engineers, electricians, or people who do wiring, such as overhead lines, cable harnesses, or circuit assemblies. Others, such as office workers and salespeople, work with it indirectly. As a source of power, electricity is accepted without much thought to the hazards encountered. Perhaps because it has become such a familiar part of our surroundings, often it is not treated with the respect it deserves.

The Bureau of Labor Statistics reports that for 1984, 3,740 work-connected deaths occurred in private-sector workplaces employing 11 workers or more. The total number of job-related injuries for that same period was roughly 5.4 million. Ten percent of the fatalities, or around *370 deaths*, were the direct result of *electrocutions at work*. What makes these statistics more tragic is that, for the most part, these accidents and fatalities could easily have been avoided.

Electricity travels in closed circuits, and its normal route is through a conductor. Shock occurs when the body becomes a part of the electrical circuit. The current must enter the body at one point and leave at another. Shock normally occurs in one of three ways. The person must come in contact with both wires of the electrical circuit; one wire of an energized circuit and the ground; or a metallic part that has become "hot" by being in contact with an energized wire, while the person is also in contact with the ground.

The metal parts of electrical tools and machines may become "hot" if there is a break in the insulation of the tool or machine wiring. The worker using these tools and machines is made less vulnerable to electrical shock when a low-resistance path from the metallic case of the tool or machine to the ground is established. This is done through the use of an equipment-grounding conductor — a low-resistance wire that causes the unwanted current to pass directly to the ground rather than through the body of the person in contact with the tool or machine. If the equipment grounding conductor has been properly installed, it has a low resistance to ground, and the worker is being protected.

The severity of the shock received when a person becomes a part of an electrical circuit is affected by three primary factors: (a) the amount of current flowing through the body (measured in amperes); (b) the path of the current through the body; and (c) the length of time the body is in the circuit. Other factors which may affect the severity of shock are the frequency of the current, the phase of the heart cycle when shock occurs, and the general health of the person prior to shock.

The effects from electric shock depend upon the type of circuit, its voltage, resistance, amperage, pathway through the body, and duration of the contact. Effects can range from a barely perceptible tingle to immediate cardiac arrest. Although there are no absolute limits of even known values which show the exact injury from any given amperage, Table 2.1 shows the general relationship between the degree of injury and amount of amperage.

In addition to shock and burn hazards, electricity poses other dangers. For example, when a short circuit occurs, hazards are created from the resulting arcs. If high current is involved, these arcs can cause injury or start a fire. Extremely high-energy arcs can damage equipment, causing fragmented metal to fly in all directions. Even low-energy arcs can cause violent explosions in atmospheres which contains explosive gases, vapors, or combustible dusts.

Electrical accidents appear to be caused by a combination of three possible factors — unsafe equipment and/or installation, workplaces made unsafe by the environment, and unsafe work practices by employees. There are various ways of protecting people from the hazards caused by electricity. These include *insulation, guarding, grounding, mechanical devices,* and *safe work practices.*

3. locating components on a suitable balcony, gallery, or platform elevated and arranged to exclude unqualified persons;
4. elevating components at 8 feet or more above the floor in some type of enclosure, such as a tray.

Entrances to rooms and other guarded locations containing exposed live parts must be marked with conspicuous warning signs forbidding unauthorized persons to enter.

Indoor electric installations serviced with over 600 volts and that are open to unqualified persons must be made with metal-enclosed equipment or enclosed in a vault or area controlled by a lock. In addition, equipment must be marked with appropriate caution signs.

Grounding is another method of protecting employees from electric shock; however, it is normally a secondary protective measure. The term "ground" refers to a conductive body, usually the earth. Used as a noun, the term means a conductive connection, whether intentional or accidental, by which an electrical circuit or equipment is connected to earth or ground plane. By "grounding" a tool or electrical system, a low-resistance path to the earth through a ground connection or connections has been intentionally created. When properly done, this path offers sufficiently low resistance and has sufficient current-carrying capacity to prevent the buildup of voltages which may result in a personnel hazard. It reduces substantially the possibilities of accidents—especially when used in combination with the other safety measures discussed previously.

There are two kinds of grounds required by "Design Safety Standards for Electrical Systems" (Subpart S). One of these is called the "services or system ground." In this instance, one wire (called "the neutral conductor" or "grounded conductor") is grounded. In an ordinary low-voltage circuit, the white (or gray) wire is grounded at the generator or transformer and again at the service entrance of the building. This type of ground is primarily designed to protect machines, tools, and insulation against damage.

To offer enhanced protection to the workers themselves, an additional ground, sometimes called the "equipment ground," must be furnished by providing another path from the tool or machine through which the current can flow to the ground. This additional ground safeguards the electrical equipment operator in the event that a malfunction causes the metal frame of the tool to become accidentally energized. The resulting heavy surge of current will then activate the circuit protection and open the circuit.

CONTROLLING ELECTRICAL HAZARDS

One way to safeguard individuals from electrically energized wires and parts is through *insulation*. An insulator is any material with high resistance to electrical current. Insulators, such as glass, mica, rubber, and plastic, are put on conductors to prevent shock, fires, and short circuits. Before employees prepare to work with electrical equipment, it is always good practice for them to check the insulation before making a connection to a power source to be sure there are no exposed wires. The insulation of flexible cords, such as extension cords, is particularly vulnerable to damage or deterioration.

The insulation that covers conductors is regulated by Subpart S or 29 CFR Part 1910, "Design Safety Standards for Electrical Systems." (This standard revises the former Subpart S and places relevant requirements of the National Electrical Code, NEC, directly into the text of the regulations, making it unnecessary for employers to refer to the NEC to determine their obligations and unnecessary for OSHA to continue incorporating the NEC by reference.)

The standard generally requires that circuit conductors, the material through which current flows, be insulated to prevent people from coming into accidental contact with the current. Also, the insulation should be suitable for the voltage and existing conditions, such as temperature, moisture, oil, gasoline, or corrosive fumes. All these factors must be evaluated before the proper choice of insulation can be made.

Conductors and cables are marked by the manufacturer to show the maximum voltage and American Wire Gage size, the type letter of the insulation, and the manufacturer's name or trademark.

Insulation is often color coded. In general, insulated wires used as equipment-grounding conductors are either continuous green or green with yellow stripes. The grounded conductors which complete a circuit are generally covered with continuous white or natural gray-colored insulation. The ungrounded conductors, or "hot wires," may be any color other than green, white, or gray. They are often colored black or red.

Energized components of electric equipment operating at 50 volts or more must be *guarded* against accidental contact. Guarding may be accomplished by:

1. locating components in a locked room, vault, or similar enclosure accessible only to qualified persons;
2. using permanent, substantial partitions or screens to exclude unqualified persons;

3. locating components on a suitable balcony, gallery, or platform elevated and arranged to exclude unqualified persons;
4. elevating components at 8 feet or more above the floor in some type of enclosure, such as a tray.

Entrances to rooms and other guarded locations containing exposed live parts must be marked with conspicuous warning signs forbidding unauthorized persons to enter.

Indoor electric installations serviced with over 600 volts and that are open to unqualified persons must be made with metal-enclosed equipment or enclosed in a vault or area controlled by a lock. In addition, equipment must be marked with appropriate caution signs.

Grounding is another method of protecting employees from electric shock; however, it is normally a secondary protective measure. The term "ground" refers to a conductive body, usually the earth. Used as a noun, the term means a conductive connection, whether intentional or accidental, by which an electrical circuit or equipment is connected to earth or ground plane. By "grounding" a tool or electrical system, a low-resistance path to the earth through a ground connection or connections has been intentionally created. When properly done, this path offers sufficiently low resistance and has sufficient current-carrying capacity to prevent the buildup of voltages which may result in a personnel hazard. It reduces substantially the possibilities of accidents—especially when used in combination with the other safety measures discussed previously.

There are two kinds of grounds required by "Design Safety Standards for Electrical Systems" (Subpart S). One of these is called the "services or system ground." In this instance, one wire (called "the neutral conductor" or "grounded conductor") is grounded. In an ordinary low-voltage circuit, the white (or gray) wire is grounded at the generator or transformer and again at the service entrance of the building. This type of ground is primarily designed to protect machines, tools, and insulation against damage.

To offer enhanced protection to the workers themselves, an additional ground, sometimes called the "equipment ground," must be furnished by providing another path from the tool or machine through which the current can flow to the ground. This additional ground safeguards the electrical equipment operator in the event that a malfunction causes the metal frame of the tool to become accidentally energized. The resulting heavy surge of current will then activate the circuit protection and open the circuit.

CONTROLLING ELECTRICAL HAZARDS

Mechanical devices are designed to automatically limit or shut off the flow of electricity in the event of a ground-fault, or overload, or short circuit in the wiring system. Fuses, circuit breakers, and ground-fault circuit interrupters are three well-known examples of such devices.

Fuses and circuit breakers are overcurrent devices which are placed in circuits to monitor the amount of current that the circuit will carry. They automatically open or break the circuit when the amount of current flow becomes excessive and therefore unsafe. Fuses are designed to melt when too much current flows through them. Circuit breakers, on the other hand, are designed to trip open the circuit.

Fuses and circuit breakers are intended primarily for the protection of conductors and equipment. They prevent overheating of wires and components which might otherwise create hazards for operators. They also open the circuit under certain hazardous ground-fault conditions.

The ground-fault circuit interrupter (GFCI) is designed to shut off electrical power within as little as 1/40 of a second. It works by comparing the amount of current going to an electrical device against the amount of current returning from the device along the circuit conductors. The GFCI is used in high-risk areas such as wet locations and construction sites.

Employees and others working with electrical equipment need to use *safe work practices*. The include de-energizing electrical eqipment before inspecting or making repairs, using electrical tools that are in good repair, using good judgment when working near energized lines, and using appropriate protective equipment. Unqualified employees or workers should avoid overhead power lines when operating ladders, cranes, or other equipment which they may operate nearby. A minimum distance should be maintained between equipment and such power lines.

The accidental or unexpected sudden starting of electrical equipment can cause severe injury or death. Before ANY inspections or repairs are made—even on the so-called low-voltage circuits—the current should be turned off at the switch box and the switch padlocked in the "off" position. At the same time, the switch or controls of the machine or other equipment being locked out of service should be securely tagged to show which equipment or circuits are being worked on.

Maintenance employees should be qualified electricians well versed in lockout procedures. No two locks should be alike; each key should fit only one lock, and only one key should be issued to each maintenance employee. If more than one employee is repairing a piece of equipment, each should lock out the switch with his or her own lock and never permit anyone else to remove it. The maintenance worker should at all times be certain that he or she is not exposing other employees to danger.

To maximize safety, an employee should always be sure to use tools that are working properly. Tools should be inspected frequently, and, those found questionable should be removed from service and properly tagged. Tools and other equipment should be regularly maintained. Inadequate maintenance can cause equipment to deteriorate, resulting in an unsafe condition.

Employees whose occupations require them to work constantly and directly with electricity must use the personal protective equipment required for the jobs they perform. This equipment may consist of rubber insulating gloves, hoods, sleeves, matting, blankets, line hose, and industrial protective helmets.

Perhaps the single most successful defense against electrical accidents is the continuous exercising of good judgment or common sense. All employees should be thoroughly familiar with the safety procedures for their particular jobs. when working around energized lines, for example, some basic procedures are (a) have the line de-energized, (b) ensure that the line remains de-energized by using some type of lockout/tagging procedure, (c) use insulated work equipment, and (d) keep a safe distance from energized lines.

The control of electrical hazards is an important part of every safety and health program. The measures suggested in this chapter should be of help in establishing such a program of control. The responsibility for a program should be delegated to individuals who have an extensive knowledge of electricity, electrical work practices, and the appropriate OSHA standards for installation and performance.

Everyone has the right to work in a safe environment. Through cooperative efforts, employers and employees can learn to identify and eliminate or control electrical hazards.

WORKPLACE HAZARDS

This section contains information developed for the employees of the Los Alamos National Laboratory and has been extracted from the Health, Safety and Environment Division's Safety Manual.

General Electrical Safety Rules

When working with possibly energized equipment, a second person capable of helping in an emergency must be present. This is not intended to apply to hand tools or completely enclosed conventional electronic equipment.

Never handle electrical equipment when hands, feet, or body are wet or perspiring or when standing on a wet floor.

The work place should be well lit, uncluttered, and access to exists should be unobstructed.

Disconnects must be labeled showing source and load.

With high voltages, regard all floors as conductive and grounded unless covered with well maintained and dry rubber matting rated for electrical work.

Whenever possible, use only one hand when working on energized circuits or control devices.

When it is necessary to touch electrical equipment (for example, when checking for overheated motors), use the back of the hand. Thus, if accidental shock were to cause muscular contraction, you would not "freeze" to the conductor.

Make sure no part of your body, especially your head, can accidentally contact energized equipment.

Avoid wearing rings, metallic watch bands, etc., when working with electrical equipment or in vicinity of induction or dielectric heating equipment.

Wear safety glasses where sparks or arcing may occur.

Avoid storing flammable liquids near electrical equipment.

Note that some equipment interlocks disconnect the high voltage source when a cabinet door is opened but power for control circuits remains on.

Equipment connected to 110-V, 3-wire receptacles must have the line which has fuses, single-pole switches, etc., connected to the ungrounded ("hot") conductor (*not* the neutral) to avoid a condition where the equipment is shut off but 110-V remains on part of the circuit.

Some 208- and 480-volt receptacles may not be wired uniformly with respect to voltage configuration, grounding, and phase rotation; therefore, before any new or relocated equipment is plugged into such receptacles, check the receptacle, label it, and, if necessary, rewire the equipment attachment plug to match the receptacle.

Remember that high voltages are not the only electrical hazard. High current conductors, as on magnets, can produce burns if accidentally shorted.

In "massive ground" areas where a shock involving the whole body is liable to occur, double-insulated tools, and ground-fault interrupters or isolation transformers should be used.

Electrical Power and Distribution

Every motor or major component should have a disconnect installed ahead of the starter. This disconnect should be capable of being padlocked in disconnected position while the motor or component is being worked on, unless it is visible and within 50 feet of it. Closely grouped equipment may be served with one disconnect.

A disconnect shall be padlocked in the open position by the person who plans to work on a motor, etc. He shall retain the key until finished. He shall also secure to the lock a tag showing his name and the time and date.

When additional people need to keep open a circuit already disconnected, they shall place additional locks, as above.

Do not open slow-opening disconnects under load, especially if the load is inductive. If, in an emergency, it should become necessary to open a disconnect under load, a long pole kept nearby or permanently attached rope could be used. The face must be averted to avoid flash-burn of eyes. Disconnects should be labeled.

Open ends of wires disconnected from equipment shall be taped or otherwise insulated.

When operating circuit breakers,

1. use only one hand;
2. keep clear of everything except operating handles;
3. turn your face away before operating the breaker;
4. don't operate isolating switches under load;
5. before closing the breaker, determine that the equipment is in condition to be energized, that all tags have been removed by the persons who placed them, that protective devices are operative, and that all persons concerned have been notified that the circuit is to be energized.

Control and test points shall be isolated from moving or energized components.

Power Tools

Portable electric tools must be grounded or be of Underwriters Laboratory (UL) "double-insultated" construction.

Stationary power tools must be grounded.

When pig-tail, clip, or similar ground connector is used, connect it first and disconnect it last. Make the connection firmly. If a clip is to be used, used one with a strong spring and insulated handle. Do not attach to painted surfaces.

When extension lights are required in damp places or inside metal vessels, the use of a flashlight of a 6- to 12-V lamp with a stepdown isolation transformer is desirable. Any 110-V leads should be outside.

The following is a suggested periodic check for portable electric equipment:

1. Examine equipment, cord, and plug. Replace if worn or damaged. (Device should be three-wire or "double-insulated.")
2. Standing on a dry, insulated floor and away from grounded objects, turn the switch on and off to see if it works properly.
3. Watch for sparks while bringing the tool lightly against a grounded object with switch on as well as off.
4. If shocks or sparks or other defects are observed, have the tool repaired.

Electronic Equipment

Equipment, whether of commercial manufacture or not, should be furnished with a circuit diagram, operating instructions, and an explanation of the associated hazards and the operation of safety devices. Door interlocks should be installed (preferably in sight) to de-energize circuits when doors are opened. If capacitors are to be discharged, then the time constant should be shown on a sign unless it is negligibly short.

In general, chassis, etc., should be grounded and bonded together. Conductors used for such purpose should be adequate for maximum anticipated fault current. Control sections and test points should be separated and isolated from high voltage sections.

High-voltage connectors should be of the type of which the body of the connector contacts its mating part before the high-voltage conductor makes contact.

Bench tops in electronic laboratories should be nonconductive, and only a minimum of connected equipment should be on the bench tops. Bench lights should be grounded.

Components used to dissipate the charge on capacitors or capacitive circuits should contain sufficient copper to dissipate the calculated maximum energy without excessive temperature rise, which endangers soldered joints, etc. Use of a discharge resistor is advisable, but the time to discharge completely should be calculated. The length of the insulated wand should be ample for the reach and the voltages. Spring clips should not be used, except where bolted connections are impractical.

Shorting conductor wands can be of rigid plastic or dry hardwood painted with insulating varnish. Conductor must be bare copper, but transparent plastic tubing may be slipped over it. The end of the conductor should terminate in a copper hook so that it can be left hanging on a terminal of a discharged capacitor during repair work. To protect capacitors, discharge resistors should be of such value as to limit peak current to 50 to 100 A and be able to absorb the energy without damage to the resistor.

When a selenium rectifier burns out or arcs over, ventilate the area to remove fumes, particularly in the case of large rectifiers.

Measurement of filament voltage on power rectifier tubes preferably should be through permanently installed volt meters rather than portable instruments. The meters should be rated for the maximum peak inverse voltage and should be protected with a glass or plastic shield.

Metal-cased meters should only be installed on grounded panels.

Filaments on some high voltage vacuum rectifier tubes are capable of producing a fatal shock if the filament circuits are ungrounded.

Rectifier tubes operating in circuits in which peak inverse voltages are greater than approximately 16,000 produce x-rays for which shielding should be provided.

Before handling cathode-ray tubes, short terminals to outer coating and ground. Store and carry them with care to prevent breakage, and wear eye protection.

Only insulated or grounded shafts should protrude through chassis panels.

Whenever technical requirements permit, install current-limiting resistors in series with the output of power supplies.

Bleeder resistors should be installed across filter capacitors in power supplies, but, even with this precaution, capacitors should be manually discharged before maintenance or repair operations are started.

If power must be on while adjusting equipment, these precautions are recommended:

1. Use insulated test prods
2. Have another person who is cognizant of hazards and familiar with artificial respiration near you.
3. Stand on an insultating mat.

Principal high-voltage danger points are listed here:

1. Transformer terminals
2. Rectifier-tube plate caps
3. Filter capacitor terminals
4. Filter choke
5. RF tuning capacitors and coils
6. Fuse panels
7. Zero adjusting screws of meters
8. Cathode-ray tube terminals

If it is necessary to disable an interlock, tag it and establish procedures for the time the interlock is inoperative and for reactivating it.

Some fail-safe considerations are

1. Safety circuits should be designed with normally open relays; and
2. Control circuit wiring should be run so that a short circuit causes a fail-safe situation.

Warning lights should be installed on equipment to indicate that power sources are "on."

Any component which is hazardous but *appears* nonhazardous should be prominently identified.

Impulse Currents: Condensers

Large condensers should be installed in barricaded locations so that all personnel are protected from hazards of bursting condensers and cables.

Keep capacitors short-circuited when not in use. Special thought must be given during design and construction of "Marx" banks to permit shorting.

Floor areas around high voltage or impulse current-generating equipment where operators or observers are likely to stand should be covered with a suitable kind of rubber matting. Matting should be properly maintained, kept dry, and discarded if visible deteriorated.

Never put your hand on or near a condenser bank or anything attached to a condenser unless a hard "ground" wire has first been attached. In all condensers except those less than 500 J, this must be preceded by discharging through a resistive "ground."

Examples of number of joules dissipated per foot of copper conductor (approximately 400°C temperature rise):

Wire size	Joules
0000	44,200
000	35,100
00	27,800
0	22,200
2	13,900
4	8,700
6	5,475
8	3,450
10	2,160
12	1,370

Significant voltage can reappear on an unshorted capacitor, so short must be left on during servicing or storage.

Miscellaneous cautions:

1. Bulging capacitors can indicate failure.
2. Insulators can fail through "creep" or "tracking."
3. Ignitrons and spark gaps can prefire!
4. Resistors can fail! Carbon and wire-wound are least reliable; ceramic types are better, and water resistors are best, but watch water level!
5. Voltmeters monitoring capacitor banks: caution — some forms of failure can bring full bank voltage to the meters.

RF and Microwaves

Electronic equipment generating rf will induce voltages in resonant circuits which may produce burns, sparks, and hazard to ammunition, volatile liquids, and gases.

High-frequency circuits may cause burns when contacted or closely approached.

Stay clear of high-frequency fields. Rings and metal watchbands should not be worn.

Areas in which microwave power density of more than 0.01 W/cm^2 is detected or suspected should be considered hazardous and warning signs posted. Personnel exposure to lower levels should be held to a minimum. The 0.01 W/cm^2 figure is currently considered good, but is subject to revision. One-tenth of this figure could be the upper limit of exposure for the eyes, particularly around 300 Mc.

If the output of generators producing average power densities greater than 0.01 W/cm^2 has to be discharged, dummy loads, water loads, or other suitable materials should be used to absorb the energy output.

Where test procedures require free-space radiation, the radiating device should be so located that the energy beam is not directed toward personnel. Reflected beams should also be considered.

Microwaves over 3000 Mc are usually reflected or absorbed by skin and can be felt by heating of surface tissue. Between approximately 1000 to 3000 Mc they can penetrate the skin and the fat layer, subject to individual variations. Frequencies below 1000 Mc penetrate the deep tissues without subjective awareness of heating. Different parts of the body vary in susceptibility to these effects. Eyes and those organs which cannot readily dissipate heat are most vulnerable.

Where microwave-generating equipment such as klystrons and magnetrons operate, spurious x-rays are generally present. In general, power supplies, oscilloscopes, electron microscopes, etc., operating over 10 kV should be checked for x-rays with survey instruments which are sensitive in the energy range of the equipment.

Where radiated power levels are high, screening or absorptive enclosures should be used around components.

Air vents and other openings should be covered with copper screen electrically connected and tightly fastened to the enclosure.

In addition to rf hazards, be alert to high dc potentials on components.

In addition to rf hazards, be alert to high dc potentials on components.

Sharp edges or points can emit corona discharges and cause burns.

High-frequency heating equipment should incorporate the following safety features:

1. High-frequency leads should be guarded.
2. Minimum suggested clearances for nonmetallic or metallic guarding:

    2,000 V peak or 150 A rf: 2 in.
    9,000 V peak or 200 A rf: 3 in.
    18,000 V peak or 320 A rf: 5 in.
3. Minimum suggested lengths of nonconducting hose (3/8 in.) for cooling water:

    2,000 V rf peak: 20 in.
    9,000 V rf peak: 36 in.
    18,000 V rf peak: 50 in.
4. For large, heavy-duty, high-frequency furnaces, individual operating instructions should be prepared and posted, covering a complete operating cycle, emergency procedures, and principal maintenance procedures.

Outdoor Locations

Body resistance between hands and feet ranges from 1500 to 20,000 ohms. If a hand or foot is wet or when a person perspires, body resistance may be reduced to about 800 to 2000 ohms. Thus, depending on weather, the danger from electric shock may be greater outdoors than indoors; and the hazard is always great when humidity is high.

Boxes, fittings, and outdoor receptacles should be weatherproof. This also applies to overcurrent devices and cabinets with doors.

Portable electric equipment should preferably be of "double-insulated" type, or supplied through a ground fault interrupter which is enclosed in a weatherproof case; or should be grounded. In the last case, ground connections should be tested periodically for continuity and visually inspected before each use.

Disconnects and breakers shall be identified and equipped with lockout devices; lockout and tagging procedures shall be observed.

CONTROLLING ELECTRICAL HAZARDS

For manholes and handholes; there should be no bare ungrounded current carrying metal parts (110 V and over) exposed to accidental contact.

Grounds should be provided for equipment in the field, including equipment in trailers.

Where cables temporarily cross roadways or paths of vehicular traffic, they must be elevated with a suitable tripod device (preferably with vertical clearance indicated) or protected by placing them in trenches or in a channel formed by two thick boards. Protection should also be provided where cables cross paths open to foot traffic.

Two or more persons shall be present when it is necessary to work on possibly energized conductors or apparatus (110 V or over).

In electrical storms; existing rules should govern what work may continue or must cease during the storm. Groups who work outdoors but have no such rules, should formulate them and make them known to all concerned.

Equipment Design

The following safety considerations are generally applicable to the design of electrical equipment for research. The special hazards of electronic equipment require that the design, fabrication, and use of such equipment consider all possible risks to which experimenters and maintenance people are exposed, and the control of these risks, including the following items:

An adequate level of light should be provided.

There should be conspicuous visual indication of both "on" and "off" conditions of each separately operable piece of hazardous equipment.

In case of high-hazard equipment, an automatic mechanical discharging device should be provided, which functions when barriers preventing human access are broken. This function must be verifiable. Protection against the hazard of the discharge itself must be provided.

In any high-power system, a convenient discharge means with an impedance capable of limiting the current to 50 amperes should be provided.

All disconnects and breakers should be labeled clearly as to the loads they control, especially if they involve high-hazard equipment.

A generally immobilizing "Emergency-off" switch, clearly identified and within easy reach of all high-hazard equipment, should be provided. Also, this switch may be used to initiate a call for help. Resetting an "Emergency-off" switch must *not* be automatic, but must require an easily understandable overt act.

Automatic safety interlocks should be provided for all access to high-hazard equipment. Interlocks should not automatically reset when barriers are reclosed. Components within enclosures should be *labeled* with standard "110 V," "220 V," "440 V," or "High Voltage" stickers.

There should be convenient, comfortable, and dry access to all equipment.

Communication equipment for use in emergencies (e.g., fire alarm box, telephone) should be provided near any hazardous equipment and marked with its location to ensure that proper instruction can be given so that people responding to a call for help can find the site quickly.

Any component which, in its common use, is nonhazardous but in its actual use may be hazardous, must be distinctly labeled. (An example might be a copper pipe carrying high voltage or high current.)

Separate and barricade operating and test sections from high-voltage components.

Procedures for Work on Energized Equipment

The following is intended to apply to work on laboratory equipment rather than plant equipment. Recognition of the hazards associated with electrical equipment in research activities is of paramount importance in developing safety guidelines for working on energized equipment, if this cannot be avoided. Two classes of electrical hazards are recognized here.

Class I. A "Class I" electrical exposure exists when the following conditions prevail:

1. The available voltage is limited to 130 V ac or 300 V dc.
2. The stored energy available in a capacitor in inductor is less than 10 joules.

Class II. "Class II" electrical exposure exists for all situations where:

1. The available voltages exceed Class I, or
2. The stored energy available exceeds Class I, or
3. There is a 30-A arcing or short-circuit possibility.

CONTROLLING ELECTRICAL HAZARDS

Work on equipment presenting a Class II hazard, with interlocks defeated or barriers removed, may be permitted only as a last resort, after all reasonable attempts have been made with the equipment unenergized or protected.

General measures to aid in preventing accidents when working on Class II energized electrical equipment include:

1. Issue written procedures for authorizing work on hazardous equipment. (These can also serve as records of maintenance.)
2. A competent and responsible person should be designated for each piece of hazardous equipment, plus at least one back-up person who is also competent and can substitute for the responsible person.
3. The supervisory chain should be identified for normal operation and development, servicing, or testing of hazardous equipment.
4. Adequate documentation should be made available to anyone working on hazardous equipment.
5. On any type of electrical equipment, as many tests as practicable should be made in the unenergized condition, or, at most, energized with reduced hazard.
6. Remove or cover clothing, long hair, and jewelry which might cause hazardous involvement.
7. Adequate lockout/tag-out procedures must be employed.
8. A person in a hazardous position who appears to be fatigued, ill, or disturbed should be replaced by a competent back-up, or the hazardous work should be terminated.
9. Supervisors and workers should be encouraged to take the conservative choice when they are in doubt about a situation regarding safety.
10. Training sessions and drills should be conducted periodically to train personnel to cope with emergencies. Cardiopulmonary resuscitation instruction should be included.

The attitudes and habits of personnel when working on energized equipment are extremely important. As an aid to addressing these attitudes, three modes of working on potentially lethal electrical equipment are identified.

Mode 1. All operations are to be conducted with the equipment positively de-energized. All external sources of electrical energy must be disconnected by some positive action (e.g., locked-out breaker) and with all internal energy sources rendered safe. "Mode 1" is a minimal hazard situation which requires little or no supervision of the worker(s) *after* the initial protective steps are taken.

Mode 2. All manipulative operations (such as making connections or alterations to *normally energized* components *or in close proximity to* normally energized components) are to be conducted with the equipment in a positively de-energized state. Measurements and observations of equipment functions are to be conducted with the equipment energized and with normal protective barriers removed or interlocks disabled. "Mode 2" is a moderate-to-severe hazard situation, depending on the operating voltages and energy capabilities of the equipment.

Mode 3. Manipulative, measurement, and observational operations are to be conducted with the equipment fully energized and with normal protective barriers removed or interlocks disable. "Mode 3" is a severe hazard situation which should be permitted only when fully justified, and should be conducted under the closet supervision and control.

Those specific safety considerations which are dependent on the degree of electrical hazard and the mode of working on equipment are classified in Table 3.1.

1. When working on Class I equipment in confined spaces or in conditions involving massive grounds, at least two knowledgeable people (i.e., capable of helping in an emergency) must be present and in position for continuous sight and sound communication.
2. In the case of Class I-3 condition, a worker must have either a co-worker or a companion within sight or hearing.
3. For II-2 and II-3 conditions, a written work *outline, checklist,* or *operating procedure is strongly recommended.*
4. In the case of Class II-2 and II-3 conditions, at least two knowledgeable people must be in position for continuous sight and sound communication.

CONTROLLING ELECTRICAL HAZARDS

Table 3.1 Risk Reduction Chart

Degree of hazard	Mode of working		
	Mode 1	Mode 2	Mode 3
Class I	Personnel may work alone with general supervision and approval	Companion required, plus general supervision and approval	Companion required, plus general supervision and approval
Class II	General supervision and approval	At least two workers at all times, plus explicit approval by supervisor	At least two workers and safety watch, with explicit approval by two levels of supervision

5. In the case of Class II-2 conditions, the explicit approval of the worker's supervisor must be obtained for the work contemplated and risks to be taken.
6. In the case of Class II-3 conditions, a third person should be identified as "safety watch." His attention should not be substantially distracted from this assignment.
7. In the case of Class II-3 conditions, the explicit approval of at least two levels of supervision (or knowledgeable co-workers) above the workers exposed to the hazard must be required.
8. When working on equipment having a Class II hazard, instructions for use in case of emergencies should be posted.
9. On Class II, the use of personal protection devices such as face shields, safety glasses, insulating mats, body protection, etc., should be provided. (Protective eye wear should always be worn to avoid injury from sparks.)
10. Under Class II-2 and II-3 conditions, the use of warning devices such as beacons, barriers, signs, etc., to indicate the presence, location, and character of the hazard is recommended. These warning devices shall be removed when the hazard ceases.
11. Connecting or rearranging test instruments on Class B equipment must be done only when it is de-energized.

CONTROLLING ELECTRICAL HAZARDS IN CONSTRUCTION

The following information has been extracted from OSHA publication No. 3097, Electrical Standards for Construction.

Electricity has long been recognized as a serious workplace hazard, exposing employees to such dangers as electric shock, burns, electrocution, fires, and explosions.

Experts in electrical safety have traditionally looked toward the widely-used National Electrical Code (NEC) for help in the practical safeguarding of persons from these hazards. The Occupational Safety and Health Administration (OSHA) recognized the important role of the NEC in defining basic requirements for safety in electrical installations by including the entire 1971 NEC by reference in Subpart K of 29 Code of Federal Regulations Part 1926. (This document can be obtained from U.S. Depart. of Labor, OSHA, Washington, D.C. 20210.)

In the rule for electrical standards for construction (OSHA 3097), dated July 11, 1986, OSHA has updated, simplified, and clarified Subpart K 29 CFR 1926. The revisions serve these objectives:

NEC requirements which directly affect employees in construction workplaces have been placed in the text of the OSHA standard, eliminating the need for the NEC to be incorporated by reference.

Certain requirements which supplemented the NEC have been integrated in the new format.

Performance language is utilized, superfluous specifications omitted, and changes in technology accommodated.

In addition, the new standard is easier for employers and employees to use and understand. Also, the OSHA revision of the electrical standards have been made more flexible, eliminating the need for constant revision to keep pace with the NEC, which is revised every three years.

The major change in the new format is the inclusion of only the relevant NEC provisions within the body of the standard itself—making it unnecessary to continue the adoption by reference of the entire NEC where some of the detailed provisions are not directly related to employee safety. The revised

CONTROLLING ELECTRICAL HAZARDS

Subpart K is divided into four major groups plus a general definitions section, as follows:

a. Installation Safety Requirements
   [29 CFR 1926.402-1026.415]
   (Sections 29 CRF 1926.402 through 1926.408 contains installation safety requirements for electrical equipment and installations used to provide electric power and light at the jobsite. These sections apply to installations, both temporary and permanent, used on the jobsite; but they do not apply to existing permanent installations that were in place before the construction activity commenced.)
b. Safety-Related Work Practices
   [29 CFR 1926.416-1926.430]
c. Safety-Related Maintenance and Environmental Considerations
   [29 CFR 1926.431-1926.440]
d. Safety Requirements for Special Equipment
   [29 CFR 1926.441-1926.448]
e. Definitions
   [29 CFR 1926.449]

Installation Safety Requirements

Part I of the standard is the most comprehensive. For this reason, the subjects included in this part are brief and selective.

If an installation is made in accordance with the 1984 National Electrical Code, it will be considered to be in compliance with Sections 1926.403 through 1926.408, except for:

1926.404(b)(1)	Ground-fault protection for employees
1926.405(a)(2)(ii)(E)	Protection of lamps on temporary wiring
1926.405(a)(2)(ii)(F)	Suspension of temporary lights by cords
1926.405(a)(2)(ii)(G)	Portable lighting used in wet or conductive locations
1926.405(a)(2)(ii)(J)	Extension cord sets and flexible cords

Approval

The electrical conductors and equipment used by the employer must be approved.

Examination, Installation, and Use of Equipment

The employer must ensure that electrical equipment is free from recognized hazards that are likely to cause death or serious physical harm to employees. Safety of equipment must be determined by:

1. Suitability for installation and use in conformity with the provisions of the standard. Suitability of equipment for an identified purpose may be evidenced by listing, labeling, or certification for that identified purpose;
2. Mechanical strength and durability, including, for parts designed to enclose and protect other equipment, the adequacy of the protection thus provided;
3. Electrical insulation;
4. Heating effects under conditions of use;
5. Arcing effects;
6. Classification by type, size, voltage, current capacity, and specific use;
7. Other factors which contribute to the practical safe-guarding of employees using or likely to come in contact with the equipment.

Guarding

Live parts of electric equipment operating at 50 volts or more must be guarded against accidental contact. Guarding of live parts must be accomplished by:

1. Location in a cabinet, room, vault, or similar enclosure accessible only to qualified persons
2. Use of permanent, substantial partitions or screens to exclude unqualified persons
3. Location on a suitable balcony, gallery, or platform elevated and arranged to exclude unqualified persons
4. Elevation of eight feet or more above the floor

Entrance to rooms and other guarded locations containing exposed live parts must be marked with conspicuous warning signs forbidding unqualified persons to enter.

Electric installations that are over 600 volts and that are open to unqualified persons must be made with metal-enclosed equipment

CONTROLLING ELECTRICAL HAZARDS

or enclosed in a vault or area controlled by a lock. In addition, equipment must be marked with appropriate caution signs.

Overcurrent Protection

The following requirements apply to overcurrent protection of circuits rated 600 volts, nominal, or less:

1. Conductors and equipment must be protected from overcurrent in accordance with their ability to conduct current safely and the conductors must have sufficient current-carrying capacity to carry the load.
2. Overcurrent devices must not interrupt the continuity of the grounded conductor unless all conductors of the circuit are opened simultaneously, except for motor-running overload protection.
3. Overcurrent devices must be readily accessible and not located where they could create an employee safety hazard by being exposed to physical damage or located in the vicinity of easily ignitable material.
4. Fuses and circuit breakers must be so located or shielded that employees will not be burned or otherwise injured by their operation (e.g., arcing).

Grounding of Equipment Connected by Cord and Plug

Exposed noncurrent-carrying metal parts of cord- and plug-connected equipment which may become energized must be grounded if:

1. In a hazards (classified) location;
2. Operated at over 150 volts to ground, except for guarded motors and metal frames of electrically heated appliances if the appliance frames are permanently and effectively insultated from ground; or
3. The equipment is one of the types listed below, but see Item 6 for exemption:
    a. Hand-held motor-operated tools.
    b. Cord- and plug-connected equipment used in damp or wet locations or by employees standing on the ground or on metal floors or working inside metal tanks or boilers.
    c. Portable and mobile x-ray and associated equipment.
    d. Tools likely to be used in wet and/or conductive locations.

e. Portable hand lamps.

f. (Exemption] Tools likely to be used in wet and/or conductive locations need not be grounded if supplied through an isolating transformer with an ungrounded secondary of not over 50 volts. Listed or labeled portable tools and appliances protected by a system of double insulation, or its equivalent, need not be grounded. If such a system is employed, the equipment must be distinctively marked to indicate that the tool or appliance uses a system of double insulation.

Safety-Related Work Practices

Protection of Employees

The employer must not permit an employee to work near any part of an electric power circuit where the employee could contact in the course of work, unless the employee is protected against shock by de-energizing the circuit and grounding it or by guarding it effectively by insulation or other means.

Where the exact location of underground electric power lines is unknown, employees using jack hammers or hand tools which may contact a line must be provided with insulated protective gloves.

Even before work is begun, the employer must determine by inquiry, observation, or instrument where any part of an exposed or concealed energized electric power circuit is located. This is necessary because a person, tool, or machine could come into physical or electrical contact with the electric power circuit.

The employer is required to advise employees of the location of such lines, the hazards involved, and protective measures to be taken as well as the post and maintain proper warning signs.

Passageways and Open Spaces

The employer must provide barriers or other means of guarding to ensure that workspace for electrical equipment will not be used as a passageway during the time when energized parts of electrical equipment are exposed. Walkways and similar working spaces must be kept clear of electric cords.

Other standards cover load ratings, fuses, cords, and cables.

Lockout and Tagging of Circuits

Tags must be placed on controls that are to be deactivated during the course of work on energized or de-energized equipment or circuits.

CONTROLLING ELECTRICAL HAZARDS

Equipment or circuits that are de-energized must be rendered inoperative and have tags attached at all points where such equipment or circuits can be energized.

Safety-Related Maintenance and Environmental Considerations

Maintenance of Equipment

The employer must ensure that all wiring components and utilization equipment in hazardous locations are maintained in a dust-tight, dust-ignition-proof, or explosion-proof condition without loose or missing screws, gaskets, threaded connections, seals, or other impairments to a tight condition.

Environmental Deterioration of Equipment

Unless identified for use in the operating environment, no conductors or equipment can be located:

1. In damp or wet locations
2. Where exposed to gases, fumes, vapors, liquids, or other agents having a deteriorating effect on the conductors or equipment
3. Where exposed to excessive temperatures

Control equipment, utilization equipment, and busways approved for use in dry locations only, must be protected against damage from the weather during building construction.

For protection against corrosion, metal raceways, cable armor, boxes, cable sheathing, cabinets, elbows, couplings, fittings, supports, and support hardware must be of materials appropriate for the environment in which they are installed.

Safety Requirements for Special Equipment

Batteries

Batteries of the unsealed type must be located in enclosures with outside vents or in well-ventilated rooms arranged to prevent the escape of fumes, gases, or electyrolyte spray into other areas. Other provisions include:

1. Ventilation to ensure diffusion of the gases from the battery and to prevent the accumulation of an explosive mixture.
2. Racks and trays treated to make them resistant to the electrolyte.

3. Floors should be acid-resistant construction unless protected from acid accumulations.
4. Face shields, aprons, and rubber gloves for workers handling acids or batteries.
5. Facilities for quick drenching of the eyes and body—within 25 feet (7.62 m) of battery-handling areas.
6. Facilities for flushing and neutralizing spilled electrolytes and for fire protection.

Battery Charging

Battery-charging installations must be located in areas designated for that purpose. When batteries are being charged, vent caps must be maintained in functioning condition and kept in place to avoid electrolyte spray. Also, charging apparatus must be protected from damage by trucks.

Related Publications

The following OSHA publications can be obtained from the OSHA Publications Distribution Office, Room S4522, Washington, D.C. 20210.

Controlling Electrical Hazards, OSHA 3975

Ground-Fault Protection on Construction Sites, OSHA 3007

National Electrical Code, National Fire Protection Association, Quincy, MA 02269, NFPA

National Electrical Code Handbook, NFPA

Electrical Safety Requirements for Employee Workplaces, NFPA

How to Prepare for Emergencies in the Workplace, OSHA 3088

All About OSHA, OSHA 2056

CONTROLLING ELECTRICAL HAZARDS IN THE HOME

Home electrical circuits are potentially hazardous because they are hidden, lurking in walls, ceilings, floors, even in outdoor areas in many instances. However, by adapting new gadgetry to the outlets and taking care to ensure that connectors, including extension cords, are in excellent condition, exposure to shock conditions can be eliminated with a knowledge of a few basic electrical facts. Some home appliances require 220 volts, but the majority of power requirement in the home is the 110-volt circuit. The user may not need to know the difference, but awareness of the 220-volt line locations is recommended.

Fuse Boxes

Whether moving into a newly constructed house or into an older house, or even into an apartment, each resident should immediately learn the location of the fuse box or circuit breaker system, and how this juncture location operates. Electrical power enters a residence via an electric meter that monitors the electricity used at that location. Power is made available through the fuse box that protects the circuits in the residence by preventing overloading of circuits that could lead to arcing and result in a fire. The fuse or circuit breaker prevents this from happening by causing an interruption in the circuit either by melting a soft metal in the fuse or, in the case of the circuit breaker type, by tripping a switch that opens the circuit. Each fuse or circuit breaker should be tagged with the location of the circuit it serves in the dwelling, for instance, kitchen, bathroom, or bedroom. This description is important in case you have power outages, or if a certain circuit needs to be disconnected, or "killed," during repair or maintenance operations in that particular circuit. The fuse box should be located in a convenient spot for easy access and should be controlled by a lock to prevent unauthorized entry.

Wall Electrical Outlets

Each circuit in a residence is usually controlled by a fuse or circuit breaker that will overload at 30 amps. Too many appliances plugged into one circuit, especially those that require high currents, such as irons and electric heaters, could cause power outages. Arcing at the plug-in socket can take place if this happens, so appliance users should be aware of overloading possibilities.

To prevent youngsters from poking things that conduct electricity, such as a hair pin or paper clip, into wall electric recepticles that are generally located near the floor, small, inexpensive covers are available that plug into the socket and prevent any such temptation. New types of wall outlets are available to prevent shocks from short circuits. Called ground-fault circuit interrupters, they perform like a circuit breaker or fuse, but they are far more sensitive. Since they are able to detect electrical leaks, such as shorts, the shut-off circuit can be reinstated at that location once the short is located. These shock protectors are becoming standard outlets in hotel bathrooms.

Also, a new electrical outlet guard is available that protects kids from shock. Adults turn a dial at the socket that blocks access to the receptacle. Also, surge protectors plugged into electrical outlets protect appliances from abnormal jolts of electricity, such as a lightning bolt, and can prevent fire.

Switches

Currently designed wall switches are designed essentially to prevent arcing when power is switched on and off. However, older switching systems over the years develop corrosion of the copper switching surfaces. Although the switch covers are well designed to prevent access to the switch terminals, any indication of arcing, such as dark streaks on the edges of the cover, or any appreciable noise during switching, indicate that the cover should be removed and the switching parts replaced. Current technology provides remote control switching with minimal contact deterioration. These switches are similar to garage door openers whereby a pushed button sends a signal to an outlet and the electricity flows. Sound-activated switches are also available, and by a hand clap the light goes on. Some require only that a hand be waved directly in front of them. No fumbling for light switches is required in these situations. Also available at hardware or electronic specialty stores are security power-failure lights that go on when electric power fails. When plugged in the wall sockets, they maintain their charge so that the batteries provide one-and-a-half hours of lumination.

Another safety gadget available is an antenna discharge unit that prevents bolts of lightning from coming into the house through TV antennas.

Extension Cords

A variety of electrical conducting flexible extension cords are available to permit transfer of the power source to desired locations. Care should be taken to ensure integrity of the complete extending system. The plug ends should be compatible with the connecting sockets. For example, some cord plugs are designed to fit only in one orientation so that a mating socket must be of the same design. Recently constructed homes have three-prong recepticles so that a ground connection is available to provide this safety feature. A three-prong plug on the end of an extension cord is the recommended type. The important consideration, of course, is to inspect for deterioration of the insulation along the cord and at the plug ends to prevent possible shorting during use. Insulation around the prongs of the plug must be intact and show no signs of charring. Use only those cords that have a tag attached from the Underwriters Laboratory. This inspection logo (U/L) guarantees the cord's integrity.

4

The National Electrical Code 1987

The first *National Electric Code®* (NEC®) was born in a National Association of Fire Engineers meeting in Richmond, Virginia, in 1881. The 1987 edition of the NEC (NFPA 70-1987) was adopted by the National Fire Protection Association, Inc., on May 21 at its 1986 Annual Meeting in Atlanta, Georgia. It was approved by the American National Standards Institute (ANSI) on July 30, 1986.

This Code is purely advisory as far as the NFPA and ANSI are concerned but is offered for use in law and for regulatory purposes in the interest of life and property protection. According to Article 90—Introduction, the purpose of the Code is the practical safeguarding of persons and property from hazards arising from the use of electricity. The Article continues by stating that the Code contains provisions considered necessary for safety, and that compliance and proper maintenance will result in an installation essentially free from hazard but not necessarily efficient, convenient, or adquate for good service or future expansion of electrical use.

Reprinted with permission from NFPA 70-1987, *National Electrical Code*, Copyright 1987, National Fire Protection Association, Quincy, MA 02269. This reprinted material is not the complete and official position of the NFPA on the referenced subject, which is represented only by the standard in its entirety.

National Electrical Code and NEC are Registered Trademarks of the National Fire Protection Association, Inc., Quincy, MA 02269.

The Code states that hazards often occur because of overloading of wiring systems by methods of usage not in conformity with the Code. This is attributed because initial wiring did not provide for increases in the use of electricity, so that an initial adequate installation and reasonable provisions for system changes will provide for future increases in the use of electricity.

It is not the intention of the Code to be used as a design specification nor as an instruction manual for untrained persons. Neither is the National Electrical Safety Code, discussed in Chapter 5; however, the differences in the two Codes can be briefly described as follows: The NEC stresses installation and maintenance of electrical systems to protect the users and the public; the NESC states that its purpose is the practical safeguarding of persons during the installation, operation, or maintenance of electric supply and communication lines and their associated equipment. Both codes are extensive references and are available from the sources listed in Chapter 1 of this volume. The National Electric Code uses a numbering system to identify the various sections. For example, the first Chapter (No. 1) is titled "General," and contains Article 100 for Definitions. (Article 90 — Introduction is the only Article preceding Article 100.) There are no subsections for Article 100; however, for the next article in the Code's Chapter 1, which is Article 110 — Requirements for Electric Installations, subsections start at number 110-1 and are numbered as high as three numbers past the Article number. Examples of the numbering system and contents of the Code are quoted in this chapter.

SCOPE OF CODE

According to Article 90 of the Code in Section 90-2, the Code covers,

> (1) Installations of electric conductors and equipment within or on public and private buildings or other structures, including mobile homes, recreation vehicles, and floating buildings; and other premises such as yards, carnival, parking and other lots, and industrial substations; ... (2) Installations of conductors that connect to the supply of electricity; (3) Installations of other outside conductors on the premises; and (4) Installation of optical fiber cable.

This same section, 90.2, explains that the

Code does not cover: (1) Installations in ships, watercraft other than floating buildings, railway rolling stock, aircraft, or automotive vehicles other than mobile homes and recreational vehicles; (2) Installations underground in mines; (3) Installations of railways for generation, transformation, transmission, or distribution of power used exclusively for operation of rolling stock or installations used exclusively for signalling and communication purposes; (4) Installations of communication equipment under the exclusive control of communication utilities, located outdoors or in building spaces used exclusively for such installations; and (5) Installations under the exclusive control of electric utilities for the purpose of communication, or metering; or for the generation, control, transformation, transmission, and distribution of electrical energy located in buildings used exclusively by utilities for such purposes or located outdoors on property owned or leased by the utility or on public highways, streets, roads, etc., or outdoors by established rights on private property.

According to a footnote in this section of the Code,

It is the intent of this section that this Code covers all premises' wiring or wiring other than utility owned metering equipment, on the load side of the service point of building structures, or any other premises not owned or leased by the utility. Also, it is the intent that this Code cover installations in buildings used by the utility for purposes other than listed ... above, such as office buildings, warehouses, garages, machine shops, and recreational buildings which are not an integral part of a generating plant, substation, or control center.

In a Special Permission section, the Code states that

The authority having jurisdiction for enforcing this Code may grant exception for the installation of conductors and equipment, not under the exclusive control of the electric utilities and used to connect the electric supply system to the service-entrance conductors of the premises served, provided such installations are outside a building or terminate immediately inside a building wall. ...

The Enforcement section, 90.4, explains that

> The Code is intended to be suitable for manditory application by government bodies exercising legal jurisdiction over electrical installation and for use by insurance inspectors.

Section 90.6 describes "Examination of Equipment for Safety whereby

> ...examinations for safety made under standard conditions will provide a basic for approval where the record is made generally available through promulgation by organizations properly equipped and qualified for experimental testing, inspections of the run of goods at factories, and service-value determination through field inspections.

Considerable space is given in Section 90.7 for the design and planning of wire circuits. Details of Wiring Planning in this section include: (1) "Plans and specifications that provide ample space in raceways, spare raceways, and additional spaces will allow for future increases in the use of electricity," (2) "Limiting the number of circuits in a single enclosure will minimize the effects from a short-circuit or ground fault in one circuit," and (3) "... metric units of measurement are in accordance with the ... International System of Units (SI)."

REQUIREMENTS FOR ELECTRICAL INSTALLATIONS

Article 110 of Chapter 1 of the Code contains vital information on mandatory specified conditions of all electrical circuits under the Code's jurisdiction. "Shall" denotes the required and mandatory conditions.

"In judging equipment," according to Section 110.3, "considerations such as the following shall be evaluated: (1) "Suitability ...," (2) "Mechanical strength and durability ...," (3) "Wire-bending and connection space. (4) Electrical insulation. (5) Heating effects ...," (6) "Arcing effects. (7) Classification by type, size, voltage, current capacity, specific use. (8) Other factors."

Voltages throughout the Code shall be considered that at which the circuit operates.

THE NATIONAL ELECTRICAL CODE 1987

Some additional requirements for Conductors, as stated in Section 110 are that (1) "Conductors normally used to carry current shall be of copper unless otherwise provided...," (2) "Conductor sizes are expressed in American Wire Gage (AWG) or in circular mils," (3) "Only wiring methods recognized as suitable are included in this Code," (4) Equipment intended to break current at fault levels shall have sufficient interrupting rating, (5) "Unless identified for use in the operating environment, no equipment shall be located in damp or wet conditions; where exposed to gases, fumes, vapors, liquids, or other agents have a deteriorating effect on the conductors or equipment; nor where exposed to excessive temperatures."

Also contained in Article 110 is its Section 110.14 dealing with Electrical Connections. The requirements include: (1) "... devices such as pressure terminal or pressure splicing connectors and soldering lugs shall be suitable for the material of the conductor...," (2) "Terminals for more than one conductor and terminals used to connect aluminum shall be so identified...," and (3) "Conductors shall be spliced or joined with spicing devices suitable for the use or by brazing, welding, or soldering with a fusible metal or alloy."

Section 110.16 points out the space required about electrical equipment for 600 volts, nominal, or less. The Code states that "Distances shall be measured from the live parts if such are exposed or from the enclosure front or opening if such are enclosed.... In addition to the dimensions shown in Table 110-16(a), the work space shall not be less than 30 inches (762 mm) wide in front of the electrical equipment.

Table 110-16(a)   Working Clearances

Voltage to Ground, Nominal	Minimum Condition:	Clear Distance (feet)		
		1	2	3
0-150		3	3	3
151-600		3	3½	4

For SI units: one inch = 25.4 millimeters; one foot = 0.3048 meter.

The "Conditions" for Table 110-16(a), according to the Code, are as follows:

REQUIREMENTS FOR ELECTRICAL INSTALLATIONS

(1) Exposed live parts on one side and no live or grounded parts on the other side of the working space, or exposed live parts on both sides effectively guarded by suitable wood or other insulating materials. Insulated wire or insulated busbars operating at not over 300 volts shall not be considered live parts.

(2) Exposed live parts on one side and grounded parts on the other side.

(3) Exposed live parts on both sides of the work space (not guarded as provided in Condition 1) with the operator between....

Working space required by this section shall not be used for storage. When normally enclosed live parts are exposed for inspection or servicing, the working space, if in a passageway or general open space, shall be suitably guarded....

At least one entrance of sufficient area shall be provided to give access to the working space about electric equipment....

For switchboards and control panels rated 1200 amperes or more and over 6 feet (1.83 m) wide, there shall be one entrance not less than 24 inches (610 mm) wide and 6½ feet (1.98 m) high at each end....

Working space with one entrance provided shall be so located that the edge of the entrance nearest the switchboards and panelboards is the minimum clear distance given in Table 110-16(a) away from such equipment....

In all cases where there are live parts normally exposed on the front of switchboards or motor control centers, the working space in front of such equipment shall not be less than 3 feet (914 mm)...

Also, "Illumination shall be provided for all working spaces about service equipment, switchboards, panelboards, or motor control centers installed indoors."

Section 110.17 contains the requirements for guarding live parts in circuits of 600 volts, nominal, or less.

Another important section is 110.21 that specified identification, or marking, of all electrical equipment.

In Section 110.34, a table (Table 110.34(a)) lists the minimum clear working space in front of electrical equipment such as switchboards, control panels, switches, circuit breakers, motor controllers, relays, and similar equipment.

Table 110-34(a) Minimum Depth of Clear Working Space in Front of Electric Equipment

Nominal Voltage to Ground	Conditions		
	1	2	3
	(Feet)	(Feet)	(Feet)
601-2500	3	4	5
2501-9000	4	5	6
9001-25,000	5	5	9
25,001-75 kV	6	8	10
Above 75 kV	8	10	12

The "Conditions" in Table 110-34(a), according to the Code, are as follows:

Exposed live parts on one side and no live or grounded parts on the other side of the working space or exposed live parts on both sides effectively guarded by suitable wood or other insulating materials. Insulated wire or insulated busbars operating at not over 300 volts shall not be considered live parts. . . .

Exposed live parts on one side and grounded parts on the other side. Concrete, brick, or tile walls will be considered as grounded surfaces. . . .

Exposed live parts on both sides of the work space (not guarded as provided in Condition 1) with the operator between. . . .

Where switches, cutouts, or other equipment operating at 600 volts, nominal, or less, are installed in a room or enclosure where there are exposed live parts or exposed wiring operating at over 600 volts, nominal, the high-voltage equipment shall be effectively separated from the space occupied by the low-voltage equipment by a suitable partition, fence, or screen. . . .

The entrances to all buildings, rooms, or enclosures containing exposed live parts or exposed conductors operating at over 600 volts, nominal, shall be kept locked. . . .

Where the voltage exceeds 600 volts, nominal, permanent and conspicuous warning signs shall be provided, reading substantially as follows: 'Warning—High Voltage—Keep Out.' . . .

Adequate illumination shall be provided for all working spaces about electrical equipment. The lighting outlets shall be so arranged that persons changing lamps or making repairs on the lighting system will not be endangered by live parts or other equipment. The points of control shall be so located that persons are not likely to come in contact with any live part or moving part of the equipment while turning on the lights. Unguarded live parts above working space shall be maintained at elevations not less than required by Table 110-34(e).

Table 110-34(e) Elevation of Unguarded Live Parts above Working Space

Nominal Voltage Between Phases	Elevation
601-7500	8'6"
7501-35000	9'
Over 35 kV	9' + 0.37" per kV above 35

WIRING DESIGN AND PROTECTION

Chapter 2 of the NEC Code provides requirements for (1) identification of terminals; (2) grounded conductors in premises wiring systems; and (3) identification of grounded conductors. The following Articles are contained in this chapter of the Code: Article 200 — Use and Identification of Grounded Conductors; Article 210 — Branch Circuits; Article 215 — Feeders; Article 225 — Outside Branch Circuits and Feeders; Article 230 — Services; Article 240 — Overcurrent Protection; Article 250 — Grounding; Article 280 — Surge Arresters.

WIRING METHODS AND MATERIALS

Chapter 3 of the NEC covers Wiring Methods and Materials.

The provisions are not intended to apply to the conductors which form an integral part of equipment, such as motors, controllers, motor control centers, or factory-assembled control equipment.

Wiring methods specified in this chapter are used for voltages 600 volts, nominal, or less where not specifically limited in some section of the chapter. They shall be permitted for voltages over 600 volts, nominal, where specifically permitted elsewhere in the NEC Code.

The following Articles are contained in Chapter 3 of the NEC: Article 300—Wiring Methods; Article 310—Conductors for General Wiring; Article 318—Cable Trays; Article 320—Open Wiring on Insulators; Article 321—Messenger Supported Wiring; Article 324—Concealed Knob-and-Tube Wiring; Article 325—Integrated Gas Space Cable; Article 326—Medium Voltage Cable; Article 328—Flat Conductor Cable Type FCC; Article 331—Electrical Nonmetallic Tubing; Article 333—Armored Cable; Article 334—Metal-Clad-Cable; Article 336—Nonmetallic-Sheathed Cable; Article 337—Shielded Nonmetallic-Sheathed Cable; Article 338—Service-Entrance Cable; Article 339—Underground Feeder and Branch-Circuit Cable; Article 340—Power and Control Tray Cable; Article 342—Nonmetallic Extensions; Article 344—Underplaster Extensions; Article 345—Intermediate Metal Conduit; Article 346—Rigid Metal Conduit; Article 347—Rigid Nonmetallic Conduit; Article 348—Electrical Metallic Tubing; Article 349—Flexible Metal Conduit; Article 351—Liquidtight Flexible Metal Conduit and Liquidtight Flexible Nonmetallic Conduit; Article 352—Surface Metal Raceways and Surface Nonmetallic Raceways; Article 353—Multioutlet Assembly; Article 354—Underfloor Raceways; Article 356—Cellular Metal Floor Raceways; Article 358—Cellular Concrete Floor Raceways; Article 362—Wireways; Article 363—Flat Cable Assemblies; Article 364—Busways; Article 365—Cablebus; Article 370—Outlet, Device, Pull and Junction Boxes, Conduit Bodies and Fittings; Article 373—Cabinets and Cutout Boxes; Article 374—Auxiliary Gutters; Article 380—Switches; Article 384—Switchboards and Panelboards.

EQUIPMENT FOR GENERAL USE

Chapter 4 of the NEC is titled, "Equipment for General Use." This Code chapter covers general requirements, applications, and construction specifications for flexible cords and flexible cables under Article 400-Flexible Cords and Cables. Limited extractions from this Article follow. Flexible cords and flexible cables shall comply with this Article and with the applicable provisions of other articles of the Code. Flexible cords and cables and their associated fittings shall be suitable for the conditions of use and location. Flexible cords and flexible cables shall conform to the description in the Code's Table 400-4 (not included here). Types of flexible cords and flexible cables other than those listed in the table shall be the subject of special investigation.

The Code states that certain popular types

> ... shall be permitted in lengths not exceeding 8 Feet (2.44 m) when attached directly, or by means of a special type of plug, to a portable appliance rated at 50 watts or less and of such nature that extreme flexibility of the cord is essential.

Rubber-filled or varnished cambric tapes shall be permitted as a substitute for the inner braids.

Certain heavier types shall be permitted for use on theater stages, in garages, and elsewhere where flexible cords are permitted by the Code.

Elevator traveling cables for operating control and signal circuits shall contain nonmetallic fillers as necessary to maintain concentricity. Cables exceeding 100 feet (30.5 m) between supports shall have steel supporting members. In locations subject to excessive moisture or corrosive vapors or gases, supporting members of other materials shall be permitted. Where steel supporting members are used, they shall run straight through the center of the cable assembly and shall not be cabled with the copper strains of any conductor.

A third conductor in these cables is for grounding purposes only.

The individual conductors of all cords, except those of heat-resistant cords, shall have a thermoset or thermoplastic insulation, except that the grounding conductor where used shall be in accordance with the Code. Unvulcanized rubber compounds shall be permitted to be used for all sizes of heater cord.

Where the voltage between any two conductors exceeds 300 but does not exceed 600, flexible cord of No. 10 and smaller shall have thermoset or thermoplastic insulation on the individual conductors at least 45 mils in thickness.

The following Articles are included in Chapter 4: Article 400—Flexible Cords and Cables; Article 402—Fixture Wires; Article 410—Lighting Fixtures, Lampholders, Lamps, Receptacles, and Rosettes; 422—Appliances; Article 424—Fixed Electric Space

Heating Equipment; Article 426—Fixed Outdoor Electric De-icing and Snow-Melting Equipment; Article 427—Fixed Electric Heating Equipment for Pipelines and Vessels; Article 430—Motors, Motor Circuits, and Controllers; Article 440—Air-Conditioning and Refrigerating Equipment; Article 445—Generators; Article 450—Transformers and Transformer Vaults; Article 460—Capacitors; Article 470—Resistors and Reactors; Article 480—Storage Batteries.

SPECIAL OCCUPANCIES

Articles 500 through 503 of Chapter 5, "Special Occupancies," of the NEC cover the requirements for electrical equipment and wiring for all voltages in locations where fire or explosion hazards may exist due to flammable gases or vapors, flammable liquids, combustible dust, or ignitible fibers or flyings.

Locations are classified depending on the properties of the flammable vapors, liquids, or gases, or combustible dusts or fibers which may be present and the likelihood that a flammable or combustible concentration or quantity is present. Where pyrophoric materials are the only materials used or handled, these locations shall not be classified.

Each room, section, or area shall be considered individually in determining its classification.

Chapter 5 of the NEC contains the following Articles: Article 500—Hazardous (Classified) Locations; Article 501—Class I Locations; Article 502—Class II Locations; Article 503—Class III Locations; Article 510—Hazardous (Classified) Location—Specific; Article 511—Commercial Garages, Repair and Storage; Article 513—Aircraft Hangars; Article 514—Gasoline Dispensing and Service Stations; Article 516—Spray Application, Dipping and Coating Processes; Article 517—Health Care Facilities; Article 518—Places of Assembly; Article 520—Theaters and Similar Locations; Article 530—Motion Picture and Television Studios and Similar Locations; Article 540—Motion Picture Projectors; Article 545—Manufactured Building; Article 547—Agricultural Buildings; Article 550—Mobile Homes and Mobile Home Parks; Article 551—Recreational Vehicles and Recreational Vehicle Parks; Article 553—Floating Buildings; Article 555—Marinas and Boatyards.

SPECIAL EQUIPMENT

Chapter 6, "Special Equipment of the NEC," covers the installation of conductors and equipment for electric signs and outline lighting in Article 600. Each outline lighting installation, and each sign of other than the portable type, shall be controlled, according to section 600-2, by an externally operable switch or breaker which will open all ungrounded conductors. The disconnecting means shall be within sight of the sign or outline lighting which it controls.

Additional extractions from Article 600 are typical of the NEC's wording.

> Switches, flashers, and similar devices controlling transformers shall be either rated for controlling inductive load(s) or have an ampere rating not less than twice the ampere rating of the transformer.
>
> The wiring method used to supply signs and outline lighting shall terminate in the sign or transformer enclosures.
>
> Every electric sign of any type, fixed or portable, shall be listed and installed in conformance with that listing, unless otherwise permitted by special permission.
>
> Signs, troughs, tube terminal boxes, and other metal frames shall be grounded.
>
> Circuits which supply lamps, ballasts, and transformers, or combinations, shall be rated not to exceed 20 amperes. Circuits containing electric-discharge lighting transformers exclusively shall not be rated in excess of 30 amperes.
>
> Each commercial building and each commercial occupancy with ground floor footage accessible to pedestrians shall be provided at an accessible location outside the occupancy, with at least one outlet for sign or outline lighting use. The outlet(s) shall be supplied by a 20-ampere branch circuit which supplies no other load....
>
> The load for the required branch circuit installed for the supply of exterior signs or outline lighting shall be computed at a minimum of 1200 volt-amperes.
>
> Signs shall be marked with the maker's name; and, for incandescent lamp signs, with the number of lampholders; and, for electric-discharge-lamp signs, with input amperes at full load and input voltage. The marking of the sign shall be visible after installation.

Transformers shall be marked with the maker's name; and transformers for electric-discharge-lamp signs shall be marked with the input rating in amperes or volt-amperes, the input voltage, and the open-circuit output voltage.

Conductors and terminals in sign boxes, cabinets, and outline troughs shall be enclosed in metal or other noncombustible material. ...

Cutouts, flashers, and similar devices shall be enclosed in metal boxes, the doors of which shall be arranged so they can be opened without removing obstructions or finished parts of the enclosure.

Enclosures shall have ample strength and rigidity. Signs and outline lighting shall be constructed of metal or other noncombustible material. Wood shall be permitted for external decoration if placed not less than 2 inches (50.8 mm) from the nearest lampholder or current-carrying part. ...

All steel parts of enclosures shall be galvanized or otherwise protected from corrosion. Enclosures for outdoor use shall be weatherproof and shall have at least two drain holes, each not larger than ½ inch (12.7 mm) or smaller than ¼ inch (6.35 mm).

Portable signs, letters, fixtures, symbols, and similar displays used in conjunction with fixed outdoor signs shall only be used when in compliance with all applicable provisions of this Code and, in addition, shall meet all of the following requirements.

A weatherproof receptacle and attachment plug having one pole for grounding shall be provided for each individual letter, fixture, or sign.

All cords shall be with one conductor grounded.

No cord shall be less than 10 feet (3.05 m) from the ground level directly underneath.

Signs and outline system enclosures shall have not less than the vertical and horizontal clearances from open conductors. The bottom of sign and outline lighting enclosures shall not be less than 16 feet (4.88 m) above areas accessible to vehicles.

The following Articles are contained in Chapter 6 of the NEC: Article 600—Electric Signs and Outline Lighting; Article 604—Manufactured Wiring Systems; Article 605—Office Furnishings; Article 610—Cranes and Hoists; Article 620—Elevators, Dumbwaiters, Escalators, and Moving Walks; Article 630—Electric

Welders; Article 640—Sound-Recording and Similar Equipment; Article 645—Electronic Computer/Data Processing Equipment; Article 650—Organs; Article 660—X-Ray Equipment; Article 665—Induction and Dielectric Heating Equipment; Article 668—Electrolytic Cells; Article 669—Electroplating; Article 670—Industrial Machinery; Article 675—Electrically Driven or Controlled Irrigation Machines; Article 680—Swimming Pools, Fountains, and Similar Installations; Article 685—Integrated Electrical Systems; Article 690—Solar Photovoltaic Systems.

SPECIAL CONDITIONS

The provisions of Article 700, Chapter 7, "Special Conditions of the NEC," apply to the electrical safety of the design, installation, operation, and maintenance of emergency systems consisting of circuits and equipment intended to supply, distribute, and control electricity for illumination and/or power to required facilities when the normal electrical supply or system is interrupted.

The Code states that

> Emergency systems are those systems legally required and classed as emergency by municipal, state, federal, or other codes, or by any governmental agency having jurisdiction. These systems are intended to supply illumination and/or power automatically to designated areas and equipment in the event of failure of the normal supply or in the event of accident to elements of a system intended to supply, distribute, and control power and illumination essential for safety to human life.

According to the Code,

> An emergency system shall have adequate capacity and rating for all loads to be operated simultaneously, and the alternate power source shall be permitted to supply emergency, legally required standby, and optional standby system loads where automatic selective load pickup and load shedding is provided as needed to assure adequate power to (1) the emergency circuits; (2) the legally required standby circuits; and (3) the optional standby circuits, in that order of priority. The alternate power source shall be permitted to be used for peak load shaving providing the above conditions are met.

A portable or temporary alternate source shall be available whenever the emergency generator is out of service for major maintenance or repair.

Section 700-7, "Signals" dictates that

Audible and visual signal devices shall be provided, where practicable, for the following purposes:

To indicate derangement of the emergency source.

To indicate that the battery is carrying load.

To indicate that the battery charger is not functioning.

To indicate a ground fault in solidly grounded emergency systems of more than 150 volts to ground and circuit protective devices rated 1000 amperes or more. The sensor for the ground-fault signal devices shall be located at, or ahead of, the main system disconnecting means for the emergency source, and the maximum setting of the signal devices shall be for a ground-fault current of 1200 amperes. Instructions on the course of action to be taken in event of indicated ground fault shall be located at or near the sensor location.

Another requirement is that a sign shall be placed at the service entrance equipment indicating type and location of on-site emergency power sources.

Chapter 7, "Special Conditions," contains the following Articles: Article 700—Emergency Systems; Article 701—Legally Required Standby Systems; Article 702—Optional Standby Systems; Article 705—Interconnected Electric Power Production Sources; Article 710—Over 600 Volts, Nominal General; Article 720—Circuits and Equipment Operating at Less than 50 Volts; Article 725—Class 1, Class 2, and Class 3 Remote-Signaling, and Power-limited Circuits; Article 760—Fire Protective Signalling Systems; Article 770—Optical Fiber Cables; Article 780—Closed-Loop and Programmed Power Distribution.

COMMUNICATIONS SYSTEMS

Article 800 of Chapter 8, "Communication Systems, of the NEC" covers telephone, telegraph (except radio), district messenger, outside wiring for fire alarm and burglar alarms, and similar central station systems; and telephone systems not connected to a central station system but using similar types of equipment, methods of installation, and maintenances.

According to the Code,

A listed protector shall be provided on each circuit run partly or entirely in aerial wire or aerial cable not confined within a block. Also, a listed protector shall be provided on each circuit, aerial or underground, so located within the block containing the building served as to be exposed to accidental contact with electric light or power conductors operating at over 300 volts to ground. (The word "block" means a square or portion of a city, town, or village enclosed by streets and including the alleys so enclosed but not any street. The word "exposed" means that the circuit is in such a position that, in case of failure of supports or insulation, contact with another circuit may result.)

Further, the Code explains:

On a circuit not exposed to accidental contact with power conductors, providing a listed protector will help protect against other hazards such as lightning and above-normal voltages induced by fault currents on power circuits in proximity to the communication circuit. When protecting a circuit run within the block between two buildings on the same premises, providing a listed protector on each end of the circuit affords protection for both buildings.

The protector shall be located in, on, or immediately adjacent to the structure or building served and as close as practicable to the point at which the exposed conductors enter or attach. The point at which the exposed conductors enter shall be considered to be the point of emergence through an exterior wall, a concrete floor slab, or from a rigid metal conduit or an intermediate metal conduit grounded to an electrode. Selecting a protector location to achieve the shortest practical protector grounding conductor will help limit potential differences between communication circuits and other metallic systems.

The following Articles are contained in Chapter 8 of the NEC: Article 800 — Communication Circuits; Article 810 — Radio and Television

Equipment; Article 820 — Community Antenna Television and Radio Distribution Systems.

TABLES AND EXAMPLES

Twenty-three pages of the NEC, Chapter 9, are devoted to data in the form of tables and examples useful to the user of electrical circuits in the myriad of ways included in the Code.

5
National Electrical Safety Code

The difference between the *National Electrical Code* (NEC) and the *National Electrical Safety Code* (NESC) can be compared to the hardware and software of a computer system. The NEC, as discussed in Chapter 4, specifies how the hardware of electrical circuits should be assembled and maintained, and, as will be presented in this chapter, the NESC "...covers the basic provisions for safeguarding of persons from hazards arising from the installation, operation, or maintenance of...[electric currents]."

The NASC is actually an American National Standard, identified as ANSI C2-1987. As explained in this Code (see Ref. 1, Chap. 1), an American National Standard implies a consensus of those substantially concerned with its scope and provisions. An American National Standard is intended as a guide to aid the manufacturer, the consumer, and the general public. The existence of an American National Standard does not in any respect preclude anyone, whether he has approved the standard or not, from manufacturing, marketing, purchasing, or using products, processes, or procedures not conforming to the standard. American National Standards are

---

Portions of ANSI C2-1987 National Electrical Safety Code © 1987 by the Institute of Electrical and Electronics Engineers, Inc., reprinted by permission of the IEEE Standards Department.

subject to periodic review and users are cautioned to obtain the latest editions.

This American National Standard may be revised or withdrawn at any time. The policy of the American National Standards Institute requires that action be taken to reaffirm, revise, or withdraw this standard no later than five years from the date of publication.

This Standard (ANSI C2-1987) covers basic provisions for safeguarding of persons from hazards arising from the installation, operation, or maintenance of (a) conductors and equipment in electric supply stations, and (b) overhead and underground electric supply and communication lines. It also includes work rules for the construction, maintenance, and operation of electric supply and communication lines and equipment.

The standard is applicable to the systems and equipment operated by utilities, or similar systems and equipment, of an industrial establishment or complex under the control of qualified persons.

This standard consists of the introduction, definitions, grounding rules, list of referenced documents, and Parts 1, 2, 3, and 4 of the 1987 edition of the National Electrical Safety Code.

This standard consists of the parts of the National Electrical Safety Code (NESC) currently in effect. The former practice of designating parts by editions has become impractical. In the 1977 edition, Parts 1 and 4 were 6th editions, Part 2 was a 7th edition, Part 3, a revision of the 6th edition, Part 2, Section 29, did not cover the same subject matter as the 5th edition, Part 3 (withdrawn in 1970). In the 1987 edition, revisions were made in all parts. It is recommended that reference to the National Electrical Safety Code be made solely by the year on the published volume and desired part number. Separate copies of the individual parts are not available.

Work on the National Electrical Safety Code started in 1913 at the National Bureau of Standards, resulting in the publication of NBS Circular 49. The last complete code (the 5th edition, MBS Handbook H30) was issued in 1948, although separate portions had been available at various times starting in 1938. In 1970, the C2 committee decided to delete from future editions Rules for the Installation and Maintenance of Electric Utilization Equipment (Part 3 of the 5th edition) now largely covered by the National Electrical Code (ANSI/NFPA 70). The 1981 edition included major changes in Parts 1, 2,

NATIONAL ELECTRICAL SAFETY CODE

and 3, minor changes in Part 4, and the incorporation of the rules pertinent to all parts into Section 1. The 1984 edition was revised to update all references and list them in a new Section 3.

The 1987 edition of the National Electrical Safety Code has been revised extensively.

The Institute of Electrical and Electronics Engineers, Inc., was designated as the administrative secretariat for C2 in January 1973, assuming the functions formerly performed by the National Bureau of Standards.

Comments on the rules, and suggestions for their improvement are invited, especially from those who have experience in their practical application. In future editions, every effort will be made to improve the rules, both in the adequacy of coverage and in the clarification of requirements. Comments should be addressed to

Secretary
National Electrical Safety Code Committee
Institute of Electrical and Electronics Engineers. Inc.
345 East 47th Street
New York, NY 10017

A representative Committee on Interpretations has been established to prepare replies to requests for interpretation of the rules contained in the code. Requests for interpretation should state the rule in question as well as the conditions under which it is being applied. Interpretations are intended to clarify the intent of specific rules and are not intended to supply consulting information on the application of the code. Requests for interpretation should be addressed to

Secretary—Interpretations
National Electrical Safety Code Committee
IEEE Standards Office
345 East 47th Street
New York, NY 10017

The code as written is a voluntary standard. However, some editions and some parts of the code have been adopted with and without changes, by some state and local jurisdictional authorities. To determine the legal status of the National Electrical Safety Code in any particular state or locality within a state, the authority having jurisdiction should be contacted.

PURPOSE AND SCOPE OF NESC

The purpose of the rules of the National Electrical Safety Code is the practical safeguarding of persons during the installation, operation, or maintenance of electric supply and communication lines and their associated equipment. They contain minimum provisions considered necessary for the safety of employees and the public. They are not intended as a design specification or an instruction manual.

These rules cover supply and communications lines, equipment and associated work practices employed by an electric supply, communication, railway, or similar utility in the exercise of its function as a utility. They cover similar systems under the control of qualified persons, such as those associated with an industrial complex.

They do not cover installations in mines, ships, railway rolling equipment, aircraft, or automotive equipment, or utilization wiring except as covered in Parts 1 and 3.

All electric supply and communication lines and equipment shall be designed, constructed, and maintained to meet the requirements of these rules. For all particulars not specified in these rules, construction and maintenance should be done in accordance with accepted good practice for the given local conditions.

These rules shall apply to all new installations and extensions, except that they may be waived or modified by the administrative authority. When so waived or modified, equivalent safety shall be provided in other ways, including special work methods.

Types of construction and methods of installation other than those specified in the rules may be used experimentally to obtain information, if done where qualified supervision is provided.

Existing installations including maintenance replacements, which comply with prior editions of the code, need not be modified to comply with these rules except as may be required for safety reasons by the administrative authority.

Where conductors or equipment are added, altered, or replaced on an existing structure, the structure or the facilities on the structure need not be modified or replaced if the resulting installation will be in compliance with the rules which were in effect at the time of the original installation.

The person responsible for an installation may modify or waive certain rules in the case of emergency or temporary installations. When the rules are waived or modified during emergencies, the installation shall be brought into compliance with these rules after the emergency has ceased. In the case of nonemergency temporary installations, only those rules involving permanence or durability of the installation may be modified.

NATIONAL ELECTRICAL SAFETY CODE

Rules in this code which are to be regarded as mandatory are characterized by the use of the word *shall*. Where a rule is of an advisory nature, to be followed insofar as practical, it is indicated by the use of the word *should*. RECOMMENDATIONS contained herein are practices considered desirable but not intended to be mandatory. NOTES contained herein, other than footnotes to tables, are for information purposes only and are not to be considered as mandatory or as part of the code requirements.

Numerical values in the requirements of this code are stated in the customary inch-foot-pound system and in the metric system (SI). In text the customary inch-foot-pound system value is shown first with the metric value (inside parentheses) following.

The metric values are *not identical equivalents* to the customary inch-foot-pound values, but have been rounded to provide convenient numbers. They are shown for information purposes only. The customary inch-foot-pound units shall govern.

Grounding Methods

Special attention is given in the introductory sections of the NESC to provide practical methods of grounding, as one of the means of safeguarding employees and the public from injury that may be caused by electrical potential. Described are methods of protective grounding of supply and communication conductors and equipment.

The rules do not cover the grounded return of electric railways nor those lightning protection wires which are normally independent of supply or communication wires or equipment.

The concern for appropriate grounding also includes the requirements for the following:

Grounding electrodes
Method of connection to electrode
Ground resistance
Separation of grounding conductors
Grounding methods for telephone and communication apparatus

Point of Connection of Grounding Conductor

Direct Current Systems Which Are to Be Grounded

750 Volts and Below Connection shall be made only at supply stations. In three-wire direct-current systems, the connection shall be made to the neutral.

Over 750 Volts Connection shall be made at both the supply and load stations. The connection shall be made to the neutral of the system. The ground or grounding electrode may be external to or remotely located from each of the stations.

One of the two stations may have its ground connection made through surge arresters provided the other station neutral is effectively grounded as described above.

Alternating Current Systems Which Are to Be Grounded

750 Volts and Below The point of the grounding connection on a wye-connected three-phase four-wire system, or on a single-phase three-wire system, shall be the neutral conductor. On other one-, two-, or three-phase systems with an associated lighting circuit or circuits, the point of grounding connection shall be on the common circuit conductor associated with the lighting circuits.

The point of grounding connection on a three-phase three-wire system, whether derived from a delta-connected or an ungrounded wye-connected transformer installation not used for lighting, may be any of the circuit conductors, or it may be a separately derived neutral.

The grounding connections shall be made at the source and at the line side of all service equipment.

Over 750 Volts (a) Nonshielded (Bare or Covered Conductors or Insulated Nonshielded Cables) Grounding connection shall be made at the neutral of the source. Additional connections may be made if desired, along the length of the neutral, where this is one of the system conductors. (b) Shielded

1. Surge-Arrester Cable-Shielding Interconnection
   Cable shielding grounds shall be bonded to surge arrester grounds, where provided, at points where underground cables are connected to overhead lines.
2. Cable Without Insulating Jacket
   Connection shall be made to the neutral of the source transformer and at cable termination points.
3. Cable with Insulating Jacket
   Additional bonding and connections between the cable insulation shielding or sheaths and the system ground are recommended. In multigrounded shielded cable systems, the shielding (including sheath) shall be

grounded at each cable joint exposed to personnel contact. Where multi-grounded shielding cannot be used for electrolysis or sheath current reasons, the shielding sheaths and splice enclosure devices shall be insulated for the voltage that may appear on them during normal operation.

Bonding transformers or reactors may be substituted for direct ground connection at one end of the cable.

Separate Grounding Conductor If a separate grounding conductor is used as an adjunct to a cable run underground, it shall be connected either directly or through the system neutral to the source transformer and accessories and cable accessories where these are to be grounded. This grounding conductor shall be located in the same direct burial or duct bank run (or the same duct if this is of magnetic material) as the circuit conductors. EXCEPTION: The grounding conductor for a circuit which is installed in a magnetic duct need not be in the same duct if the duct containing the circuit is bonded to the separate grounding conductor at both ends.

Messenger Wires and Guys

Messenger Wires Messenger wires required to be grounded shall be connected to grounding conductors at poles or structures at maximum intervals as listed below.

a. Where messenger wires are adequate for system grounding conductors, four connections in each mile (1.6 km).

b. Where messenger wires are not adequate for system grounding conductors, eight connections per mile (1.6 km), exclusive of service grounds.

Guys Guys which are required to be grounded shall be connected to one or more of the following:

a. A grounded metallic supporting structure.

b. An effective ground on a nonmetallic supporting structure.

c. A line conductor which has at least four ground connections in each mile of line in addition to the ground connections at individual services.

Common Grounding of Messengers and Guys on the Same Supporting Structure Where messengers and guys on the same supporting structure are required to be grounded, they shall be connected to the same grounding conductor or to separate grounding conductors that are bonded together at common crossing structures and at intervals specified in this Code. EXCEPTION: This rule does not apply to guys that are connected to an effectively grounded overhead static wire.

Current in Grounding Conductor

Ground connection points shall be so arranged that under normal circumstances there will be no objectionable flow of current over the grounding conductor. If an objectionable flow of current occurs over a grounding conductor due to the use of multiple grounds, one or more of the following should be used:

1. Abandon one or more grounds.
2. Change location of grounds.
3. Interrupt the continuity of the conductor between ground connections.
4. Subject to the approval of the administrative authority take other effective means to limit the current.

The system ground of the source transformer shall not be removed.

The temporary currents set up under abnormal conditions while the grounding conductors are performing their intended protective functions are not considered objectionable. The conductor shall have the capability of conducting anticipated fault current without thermal overloading or excessive voltage buildup.

Fences

Fences, where required to be grounded by other parts of this code, shall be grounded at or near the location of a supply line or lines crossing them, and, additionally, at distances not exceeding 150 ft. (45 m) on either side. Fences shall also be grounded at each side of a gate or other opening in the fence. Any gate or other opening shall also be bonded across by a buried bonding jumper. A gate shall be metallically connected or bonded to the grounding conductor, jumper, or fence.

NATIONAL ELECTRICAL SAFETY CODE

Separate barbed wire strands above fencing, or nonconducting posts, shall be bonded to metallic fencing or grounding conductors at the grounding points.

Where required to be grounded, fences shall be bonded to the grounding system of the enclosed equipment or to a separate underground conductor below or near the fence line.

Grounding Conductor and Means of Connection

Composition of Grounding Conductors

In all cases, the grounding conductor shall be made of copper or other metals or combinations of metals which will not corrode excessively during the expected service life under the existing conditions and, if practical, shall be without joint or splice. If joints are unavoidable, they shall be so made and maintained as not to increase materially the resistance of the grounding conductor and shall have appropriate mechanical and corrosion-resistant characteristics. For surge arresters and ground detectors, the grounding conductor or conductors shall be as short, straight, and free from sharp bends as practical. The structural metal frame of a building or structure may serve as a grounding conductor to an acceptable grounding electrode.

In no case shall a circuit-opening device be inserted in the grounding conductor or connection except where its operation will result in the automatic disconnection from all sources of energy of the circuit leads connected to the equipment so grounded. EXCEPTION 1: Temporary disconnection of grounding conductors for testing purposes, under competent supervision, shall be permitted. EXCEPTION 2: Disconnection of a grounding conductor from a surge arrester is allowed when accomplished by means of a surge arrester disconnector. NOTE: The base of the surge arrester may remain at line potential following operation of the disconnector.

Connection of Grounding Conductors

Connection of the grounding conductor shall be made by a means matching the characteristics of both the grounded and grounding conductors, and suitable for the environmental exposure. These means include brazing, welding, mechanical and compression connections, ground clamps, and ground straps.

Soldering is acceptable only in conjunction with lead sheaths.

Ampacity and Strength

The short time ampacity of a bare grounding conductor is that current which the conductor can carry for the time during which the current flows without melting or separating under the applied tensions. If a grounding conductor is insulated, its short time ampacity is the current which it can carry for the applicable time without damaging the insulation. Where grounding conductors at one location are paralleled, the increased total current capacity may be considered.

System Grounding Conductors for Single-Grounded Systems The system grounding conductor or conductors for a system with single system grounding electrode or set of electrodes, exclusive of grounds at individual services, shall have a short time ampacity adequate for the fault current which can flow in the grounding conductor or conductors for the operating time of the system protective device. If this value cannot be readily determined, continuous ampacity of the grounding conductor or conductors shall be not less than the full load continuous current of the system supply transformer or other source of supply.

System Grounding Conductors for Multigrounded Alternating Current Systems The system grounding conductors for an alternating current system with grounds at more than one location exclusive of grounds at individual services shall have continuous total ampacities at each location of not less than one-fifth that of the conductors to which they are attached.

Grounding Conductors for Instrument Transformers The grounding conductor for instrument cases and secondary circuits for instrument transformers shall not be smaller than AWG No. 12 copper or shall have equivalent short time ampacity.

Grounding Conductors for Primary Surge Arresters The grounding conductor or conductors shall have adequate short time ampacity under conditions of excess current caused by or following a surge. Individual arrester grounding conductors shall be no smaller than AWG No. 6 copper or AWG No. 4 aluminum. EXCEPTION: Arrester grounding conductors may be copper-clad or aluminum-clad steel wire having not less than 30% of the conductivity of solid copper or aluminum wire of the same diameter.

Where flexibility of the grounding conductor, such as adjacent to the base of the arrester, is vital to its proper operation, a suitably flexible conductor shall be employed.

Grounding Conductors for Equipment, Messenger Wires, and Guys

a. *Conductors.* The grounding conductors for equipment, raceways, cable, messenger wires, guys, sheaths, and other metal enclosures for wires shall have short time ampacities adequate for the available fault current and operating time of the system fault protective device. if no overcurrent or fault protection is provided, the ampacity of the grounding conductor shall be determined by the design and operating conditions of the circuit but shall not be less than that of AWG No. 8 copper. Where the adequacy and continuity of the conductor enclosures and their attachment to the equipment enclosures is assured, this path can constitute the equipment grounding conductor.

b. *Connections.* Connections of the grounding conductor shall be to a suitable lug, terminal, or device not disturbed in normal inspection, maintenance, or operation.

Fences

The grounding conductor for fences required to be grounded by other parts of this code shall be any of those meeting the requirements of this Code or shall be steel wire not smaller than Stl WG No. 5. It shall be connected to the fence posts with connecting means suitable for the material when the posts are of conducting material. If the posts are of nonconducting material, suitable bonding connections shall be made to the fence mesh strands and the barbed wire strands at each grounding conductor point.

Bonding of Equipment Frames and Enclosures

Where required, a low impedance metallic path shall be provided for the passage of possible conductor or equipment, or both, fault current back to the grounded terminal of the supply, where the supply is local. Where the supply is remote, the metallic path shall interconnect the equipment frames and enclosures with all other nonenergized conducting components within reach and shall additionally be connected to ground as outlined in Rule 93C5. Short-time ampacities of bonding conductors shall be adequate for the duty involved.

Ampacity limit No grounding conductor need have greater ampacity than either:

a. The phase conductors which would supply the ground fault current, or

b. The maximum current which can flow through it to the ground electrode or electrodes to which it is attached. For single grounding conductor and connected electrode or electrodes, this would be the supply voltage divided by the electrode resistance (approximately).

Strength All grounding conductors shall have mechanical strength suitable for the conditions to which they may reasonably be subjected.

Further, unguarded grounding conductors shall have a tensile strength not less than that of AWG No. 8 softdrawn copper, except as noted in this Code.

Guarding and Protection

1. The grounding conductors for single-grounded systems and those exposed to mechanical damage shall be guarded. However, grounding conductors need not be guarded where not readily accessible to the public nor where grounding multigrounded circuits or equipment.

2. Where guarding is required, grounding conductors shall be protected by guards suitable for the exposure to which they may reasonably be subjected. The guards should extend for not less than 8 ft. (2.45 m) above the ground or platform from which the grounding conductors are accessible to the public.

3. Where guarding is not required, grounds shall be protected by being substantially attached closely to the surface of the pole or other structure in areas of exposure to mechanical damage and, where practical, on the portion of the structure having least exposure.

4. Guards used for grounding conductors of lightning protection equipment shall be nonmagnetic materials if the guard completely encloses the grounding conductor or is not bonded at both ends to the grounding conductor.

Underground

1. Grounding conductors laid directly underground shall be laid slack or shall be of sufficient strength to prevent being readily broken by earth movement or settling normal at the particular location.

2. Direct-buried uninsulated joints or splices in grounding conductors should be welded, brazed, or of the compression type to minimize the possibility of loosening or corrosion. The number of joints or splices should be the minimum practical.

3. Grounding cable insulation shielding systems shall be interconnected with all other accessible grounded power supply equipment in manholes, handholes, and vaults. EXCEPTION: Where cathodic protection or shield cross-bonding is involved, interconnection may be omitted.

4. Looped magnetic elements such as structural steel, piping, reinforcing bars, etc., should not separate grounding conductors from the phase conductors of circuits they serve.

5. Metals used for grounding, in direct contact with earth, concrete, or masonry, shall have been proven suitable for such exposure.

NOTE 1: Under present technology, aluminum has not generally been proven suitable for such use.

NOTE 2: Metals of different galvanic potentials which are electrically interconnected may require protection against galvanic corrosion.

6. Sheath transportation connections (cross-bonding):

a. Where cable insulating shields or sheaths, which are normally connected to the ground, are insulated from ground to minimize shield circulating currents, they shall be insulated from personnel contact at accessible locations. Transposition connections and bonding jumpers shall be insulated for nominal 600 volts service, unless the normal shielding voltage exceeds this level, in which case the insulation shall be ample for the working voltage to ground.

b. Bonding jumpers and connecting means shall be sized and selected to carry the available fault current without damaging jumper insulation or sheath connections.

*Common Grounding Conductor for Circuits,
Metal Raceways, and Equipment*

Where the ampacity of a supply system grounding conductor is also adequate for equipment grounding requirements, this conductor may be used for the combined purpose. Equipment referred to includes the frames and enclosures of supply system control and auxiliary components, conductor raceways, cable shields, and other enclosures.

INSTALLATION AND MAINTENANCE OF ELECTRIC SUPPLY STATIONS AND EQUIPMENT (Part 1)

The purpose of Part 1 of this Code is the practical safeguarding of persons during the installation, operation, or maintenance of electric supply stations and their associated equipment. It covers the electric supply conductors and equipment, along with the associated structural arrangements in electric supply stations, which are accessible only to qualified personnel. It also covers the conductors and equipment employed primarily for the utilization of electric power when such conductors and equipment are used by the utility in the exercise of its function as a utility.

Protective Arrangements in Electric Supply Stations

Enclosure of Equipment

Rooms and spaces in which electric supply conductors or equipment are insulated shall be so arranged with fences, screens, partitions, or walls as to minimize the possibility of entrance of unauthorized persons or interference by them with equipment inside. Entrances not under observation of an authorized attendant shall be kept locked.

Warning signs shall be displayed at entrances.

Metal fences, when used to enclose electric supply stations having energized electrical conductors or equipment, shall have a height not less than 7 ft. (2.13 m) overall and shall be grounded in accordance with this Code.

The requirements for fence height may be satisfied with any one of the following:

1. Fence fabric, 7 ft. (2.13 m) or more in height.

2. A combination of 6 ft. (1.80 m) or more of fence fabric and a 1 ft. (300 mm) extension utilizing three or more strands of barbed wire.
3. Other types of construction, such as nonmetallic material, which present equivalent barriers to climbing or other unauthorized entry.

Rooms and Spaces

All rooms and spaces in which electric supply equipment is installed shall comply with the following requirements:
Construction They shall be as much as practical noncombustible.
Use They should be as much as practical free from combustible materials, dust, and fumes and shall not be used for manufacturing or for storage, except for minor parts essential to the maintenance of the installed equipment.
Ventilation There should be sufficient ventilation to maintain operating temperatures within ratings, arranged to minimize accumulation of airborne contaminants under any operating conditions.
Moisture and Weather They should be dry. In outdoor stations or stations in wet tunnels, subways, or other moist or high humidity locations, the equipment shall be suitably designed to withstand the prevailing atmospheric conditions.

Electric Equipment

All stationary equipment shall be supported and secured in a manner consistent with reasonably expected conditions of service. Consideration shall be given to the fact that certain heavy equipment, such as transformers, can be secured in place by their weight. However, equipment that generates dynamic forces during operation may require appropriate additional measures.

Illumination

Under Normal Conditions

Rooms and spaces shall have means for artificial illumination. Illumination levels not less than those listed in Table 5.1 (pp. 104-105) are recommended for safety to be maintained on the task.

Emergency Lighting

1. A separate emergency source of illumination with automatic initiation, from an independent generator, storage battery, or other suitable source, shall be provided in every attended station.
2. Emergency lighting of 1 footcandle (11 lux) shall be provided in exit paths from all areas of attended stations. Consideration must be given to the type of service to be rendered whether of short time or long duration. The minimum duration shall be 1½ hours. It is recommended that emergency circuit wiring shall be kept independent of all other wiring and equipment.

Fixtures

Arrangements for permanent fixtures and plug receptacles shall be such that portable cords need not be brought into dangerous proximity to live or moving parts. All lighting shall be controlled and serviced from safely accessible locations.

Attachment Plugs and Receptacles for General Use

Portable conductors shall be attached to fixed wiring only through separable attachment plugs which will disconnect all poles by one operation. Receptacles installed on two or three wire single phase, ac branch circuits shall be of the grounding type. Receptacles connected to circuits having different voltages, frequencies, or types of current (ac or dc) on the same premises shall be of such design that attachment plugs used on such circuits are not interchangeable.

Receptacles in Damp or Wet Locations

All 120 V ac permanent receptacles shall either be provided with ground fault interrupter (GFI) protection or be on a grounded circuit which is tested at such intervals as experience has shown to be necessary.

Floor, Floor Openings, Passageways, Stairs

Floors

Floors shall have even surfaces and afford secure footing. Slippery floors or stairs should be provided with antislip covering.

NATIONAL ELECTRICAL SAFETY CODE

Passageways

Passageways, including stairways, shall be unobstructed and shall, where practical, provide at least 7 ft. (2.13 m) headroom. Where the preceding requirements are not practical, the obstructions should be painted, marked, or indicated by warning signs and the area properly lighted.

Railings

All floor openings without gratings or other adequate cover and raised platforms and walkways in excess of 1 ft. (300 mm) in height shall be provided with railings. Openings in railings for units such as fixed ladders, cranes, and the like shall be provided with adequate guards such as grates, chains, or sliding pipe sections.

Stair Guards

All stairways consisting of four or more risers shall be provided with handrails.

Top Rails

All top rails shall be kept unobstructed for a distance of 3 in. (75 mm) in all directions except from below at supports.

Exits

Clear Exits

Each room or space and each working space about equipment shall have a means of exit which shall be kept clear of all obstructions.

Double Exits

If the plan of the room or space and the character and arrangement of equipment are such that an accident would be likely to close or make inaccessible a single exit, a second exit shall be provided.

Exit Doors

Exit doors shall swing out and be equipped with panic bars, pressure plates, or other devices that are normally latched but open under simple pressure. EXCEPTION: This rule does not apply to exit doors in buildings and rooms containing low-voltage, nonexplosive equipment, and to gates in fences for outdoor equipment installations.

Fire Extinguishing Equipment

Fire extinguishing approved for the intended use shall be conveniently located and conspicuously marked.

Included in Part 1 are also the requirements for the following:

Installation and maintenance equipment

Rotating equipment

Storage batteries

Transformers and regulators

Conductors

Circuit breakers, reclosers, switches, fuses

Switchgear and metal enclosed bus

Surge arresters

INSTALLATION AND MAINTENANCE OF OVERHEAD ELECTRIC SUPPLY AND COMMUNICATION LINES (Part 2)

The purpose of Part 2 of this Code is the practical safeguarding of persons during the installation, operation, or maintenance of overhead supply and communication lines and their associated equipment. It covers supply and communication conductors and equipment in overhead lines. It covers the associated structural arrangements of such systems and the extension of such systems into buildings. The rules include requirements for spacing, clearances, and strength of construction. They do not cover installations in electric supply stations. Also, rules covering supply line influence and communication line susceptiveness have not been detailed in this Code. Cooperative procedures are recommended in the control of voltages induced from proximate facilities. Therefore, reasonable advance notice should be given to owners or operators of other proximate facilities which may be adversely affected by new construction or changes in existing facilities. And, all parts which must be examined or adjusted during operation shall be arranged so as to be accessible to authorized persons by the provision of adequate climbing spaces, working spaces, working facilities, and clearances between conductors.

Inspection and Tests of Lines and Equipment

When in Service

1. *Initial compliance with rules.* Lines and equipment shall comply with these safety rules when placed in service.
2. *Inspection.* Lines and equipment shall be inspected at such intervals as experience has shown to be necessary.
3. *Tests.* When considered necessary, lines and equipment shall be subjected to practical tests to determine required maintenance.
4. *Record of defects.* Any defects affecting compliance with this code revealed by inspection or tests, if not promptly corrected, shall be recorded; such records shall be maintained until the defects are corrected.
5. *Remedying defects.* Lines and equipment with recorded defects which could reasonably be expected to endanger life or property shall be promptly repaired, disconnected, or isolated.

When Out of Service

1. *Lines infrequently used.* Lines and equipment infrequently used shall be inspected or tested as necessary before being placed into service.
2. *Lines temporarily out of service.* Lines and equipment temporarily out of service shall be maintained in a safe condition.
3. *Lines permanently abandoned.* Lines and equipment permanently abandoned shall be removed or maintained in a safe condition.

Grounding of Circuits, Supporting Structures, and Equipment

Methods

Grounding required by these rules shall be in accordance with the applicable methods given in this Code.

Circuits

1. *Common neutral.* A conductor used as a common neutral for primary and secondary circuits shall be effectively grounded as specified in this Code.
2. *Other neutrals.* Primary or secondary neutral conductors, other than common neutrals, which are to be effectively grounded, shall be grounded as specified in this Code.
3. *Surge arresters.* Where the operation of surge arresters is dependent upon grounding, they shall be grounded in accordance with the methods outlined in this Code.
4. *Use of earth as part of circuit.* Supply circuits shall not be designed to use the earth normally as the sole conductor for any part of the circuit.

Noncurrent-Carrying Parts

General Metal or metal reinforced supporting structures, including lamp posts; metal conduits and raceways; cable sheaths; messengers; metal frames, cases, and hangers of equipment; and metal switch handles and operating rods shall be effectively grounded. EXCEPTION 1: This rule does not apply to frames, cases, and hangers of equipment and switch handles and operating rods which are 8 ft. (2.45 m) or more above readily accessible surfaces or are otherwise isolated or guarded and where the practice of not grounding such items has been a uniform practice over a well defined area. EXCEPTION 2: This rule does not apply to isolated or guarded equipment cases in certain specialized applications, such as series capacitors where it is necessary that equipment cases be either ungrounded or connected to the circuit. Such equipment cases shall be considered as energized and shall be suitably identified. EXCEPTION 3: This rule does not apply to equipment cases, frames, equipment hangers, conduits, raceways, and cable sheaths enclosing only communications conductors, provided they are not exposed to probable contact with open supply conductors of over 300 volts.

Guys Guys shall be effectively grounded if attached to a supporting structure carrying any supply conductor of more than 300 volts or if exposed to such conductors. EXCEPTION 1: This rule does not apply to guys containing an insulator or

NATIONAL ELECTRICAL SAFETY CODE

insulators installed in accordance with and meeting the requirements of this Code. EXCEPTION 2: This rule does not apply to guys attached to supporting structures if all supply conductors are in cable conforming to the requirements of this Code. EXCEPTION 3: This rule does not apply if the guy is attached to a supporting structure on private right-of-way if all the supply circuits exceeding 300 volts meet the requirements of this Code.

Multiple messengers on the same structure Communication cable messengers exposed to power contacts, power induction, or lightning, shall be bonded together at intervals specified in Rule 92C.

Arrangement of Switches

Accessibility

Switches or their control mechanisms shall be installed so as to be accessible to authorized persons.

Indicating Open or Closed Position

Switch position shall be visible or clearly indicated.

Locking

Switching operating mechanisms which are accessible to unauthorized persons shall have provisions for locking in each operational position.

Uniform Position

The handles or control mechanisms for all switches throughout any system should have consistent positions when opened and uniformly different positions when closed in order to minimize operating errors. Where this practice is not followed, the switches should be marked to minimize mistakes in operation.

Relations Between Various Classes of Lines

Standardization of Levels

The levels at which different classes of conductors are to be located should be standardized by agreement of the utilities concerned.

Relative Levels: Supply and Communication Conductors

Preferred levels Where supply and communication conductors cross each other or are located on the same structures, the supply conductors should be carried at the higher level. EXCEPTION: This rule does not apply to trolley feeders which may be located for convenience approximately at the level of the trolley-contact conductor.

Special construction for supply circuits, the voltage of which is 600 volts or less and carrying power not in excess of 5 kilowatts Where all circuits are owned or operated by one party or where cooperative consideration determines that the circumstances warrant and the necessary coordinating methods are employed, single-phase alternating-current or two-wire direct-current circuits carrying a voltage of 600 volts or less between conductors, with transmitted power not in excess of 5 kilowatts, when involved in the joint use of structures with communication circuits may be installed in accordance with the Code under the following conditions.

a. That such supply circuits are of covered conductor not smaller than AWG No. 8 medium hard-drawn copper or its equivalent in strength, and the construction otherwise conforms with the requirements for supply circuits of the same class.

b. That the supply circuits be placed on the end and adjacent pins of the lowest through signal support arm and that a 30 inch (750 mm) climbing space be maintained from the ground up to a point at least 24 in. (600 mm) above the supply circuits. The supply circuits shall be rendered conspicuous by the use of insulators of different form or color from others on the pole line or by stenciling the voltage on each side of the support arm between the pins carrying each supply circuit, or by indicating the voltage by means of metal characters.

c. That there shall be a vertical clearance of at least 2 ft. (600 mm) between the support arm carrying these supply circuits and the next support arm above. The other pins on the support arm carrying the supply circuits may be occupied by communication circuits used in the operation or control of signal system or other supply system if owned, operated, and maintained by the same company operating the supply circuits.

d. That such supply circuits shall be equipped with arresters and fuses installed in the supply end of the circuit and where the signal circuit is alternating current, the protection shall be installed on the secondary side of the supply transformer. The arresters shall be designed so as to break down at approximately twice the voltage between the wires of the circuit, but the breakdown voltage of the arrester need not be less than 1 kilovolt. The fuses shall have a rating not in excess of approximately twice the maximum operating current of the circuit, but their rating need not be less than 10 amperes. The fuses likewise shall in all cases hav a rating of at least 600 volts, and, where the supply transformer is a stepdown transformer, shall be capable of opening the circuit successfully in the event the transformer primary voltage is impressed upon them.

e. Such supply circuits in cable meeting the requirements of the Code may be installed below communication attachments, with not less than 2 ft. (600 mm) vertical separation between the supply cable and the lowest communication attachment. Communication circuits other than those used in connection with the operation of the supply circuits shall not be carried in the same cable with such supply circuits.

f. Where such supply conductors are carried below communication conductors, transformers and other apparatus associated therewith shall be attached only to the sides of the support arm in the space between, and at no higher level than, such supply wires.

g. Lateral runs of such supply circuits carried in a position below the communication space shall be protected through the climbing space by wood molding or equivalent covering, or shall be carried in insulated multiple-conductor cable, and such lateral runs shall be placed on the under side of the support arm.

Relative Levels: Supply Lines of Different Voltage Classifications (as Classified in the Code)

At crossings or conflicts Where supply conductors of different voltage classifications cross each other or structure conflict exists, the higher voltage lines should be carried at the higher level.

On structures used only by supply conductors Where supply conductors of different voltage classifications are on the same structures, relative levels should be as follows:

    a. Where all circuits are owned by one utility, the conductors of higher voltages should be placed above those of lower voltage.

b. Where different circuits are owned by separate utilities, the circuits of each utility may be grouped together, and one group of circuits may be placed above the other group provided that the circuits in each group are located so that those of higher voltage are at the higher levels and that any of the following conditions is met:

(1) A vertical spacing of not less than that required is maintained between the nearest line conductors of the respective utilities.

(2) Conductors of a lower voltage classification placed at a higher level than those of a higher classification shall be placed on the opposite side of the structure.

(3) Ownership and voltage are prominently displayed.

Joint Use of Structures

Joint use of structures should be considered for circuits along highways, roads, streets, and alleys. The choice between joint use of structures and separate lines shall be determined through cooperative consideration of all the factors involved, including the character of circuits, the total number and weight of conductors, tree conditions, number and location of branches and service drops, possible structure conflicts, availability of right-of-way, etc. Where such joint use is mutually agreed upon, it shall be subject to the appropriate grade of construction specified in this Code.

In addition to the requirements in Part 2 discussed above, the NESC lists stipulations for the following:

Clearances of supporting structures

Vertical clearance above ground

Clearances for conductors on different supports

Clearances from buildings, bridges, rail cars, pools

Clearances for conductors on same support

Clearances in working spaces

Clearances for vertical and lateral conductors on supports

Grades of construction

General loading requirements

Strength requirements

Line insulation.

INSTALLATION AND MAINTENANCE OF UNDERGROUND ELECTRIC SUPPLY AND COMMUNICATION LINES (Part 3)

The purpose of Part 3 of this Code is the practical safeguarding of persons during the installation, operation, or maintenance of underground or buried supply and communication cables and associated equipment. It covers supply and communication cables and equipment in underground or buried systems. The rules cover the associated structural arrangements and the extension of such systems into buildings. It also covers the cables and equipment employed primarily for the utilization of electric power when such cables and equipment are used by the utility in the exercise of its function as a utility. They do not cover installations in electric supply stations.

Installation and Maintenance

1. Persons responsible for underground facilities shall be in a position to indicate the location of their facilities.
2. Reasonable advance notice should be given to owners or operators of other proximate facilities which may be adversely affected by new construction or changes in existing facilities.
3. All parts which must be examined or adjusted during operation shall be arranged so as to be readily accessible to authorized persons by the provision of adequate working spaces, working facilities, and clearances.

Inspection and Tests of Lines and Equipment

When in Service

1. *Initial compliance with safety rules.* Lines and equipment shall comply with these safety rules upon being placed in service.
2. *Inspection.* Accessible lines and equipment shall be inspected by the responsible party at such intervals as experience has shown to be necessary.
3. *Tests.* When considered necessary, lines and equipment shall be subjected to practical tests to determine required maintenance.
4. *Record of defects.* Any defects affecting compliance

with this Code revealed by inspection, if not promptly corrected, shall be recorded; such record shall be maintained until the defects are corrected.

5. *Remedying defects.* Lines and equipment with recorded defects which would endanger life or property shall be properly repaired, disconnected, or isolated.

When Out of Service

1. *Lines infrequently used.* Lines and equipment infrequently used shall be inspected or tested as necessary before being placed into service.
2. *Lines temporarily out of service.* Lines and equipment temporarily out of service shall be maintained in a safe condition.
3. *Lines permanently abandoned.* Lines and equipment permanently abandoned shall be removed or maintained in a safe condition.

Grounding of Circuits and Equipment

Methods

The methods to be used for grounding of circuits and equipment are given in this Code.

Conductive Parts to be Grounded

Cable sheaths and shields (except conductor shields), equipment frames, and cases (including pad-mounted devices), and conductive lighting poles shall be effectively grounded. Ducts and riser guards of conductive material which enclose electric supply lines shall be effectively grounded. EXCEPTION: This rule does not apply to parts which are 8 ft. (2.45 m) or more above readily accessible surfaces or are otherwise isolated or guarded.

Use of Earth as Part of Circuit

Supply circuits shall not be designed to use the earth normally as the sole conductor for any part of the circuit.

Communication Protective Requirements

Where Required

Where communications apparatus is handled by other than qualified persons, it shall be protected by one or more of the means listed in Rule 315B if such apparatus is permanently connected to lines subject to any of the following:

1. Lightning
2. Possible contact with supply conductors whose voltage exceeds 300 V
3. Transient rise in ground potential exceeding 300 V
4. Steady-state induced voltage of a level which may cause personal injury.

NOTE: When communications cables will be in the vicinity of supply stations where large ground currents may flow, the effect of these currents on communications circuits should be evaluated.

Means of Protection

Where communications apparatus is required to be protected, protective means adequate to withstand the voltage expected to be impressed shall be provided by insulation, protected where necessary by surge arresters. Severe conditions may require the use of additional devices such as auxiliary arresters, drainage coils, neutralizing transformers, or isolating devices.

Induced Voltage

Rules covering supply line influence and communication line susceptiveness have not been detailed in this Code. Cooperative procedures are recommended to minimize steady state voltages induced from proximate facilities. Therefore, reasonable advance notice should be given to owners or operators of other known proximate facilities which may be adversely affected by new construction or changes in existing facilities.

In addition to the requirements discussed in Part 3 above, the following items are included in Part 3 of this Code:

underground conduit systems
ducts and joints
manholes, handholes, and vaults
supply cable
cable on underground structures

direct buried cable
risers
equipment underground
installation in tunnels.

OPERATION OF ELECTRIC-SUPPLY AND COMMUNICATIONS LINES AND EQUIPMENT (Part 4)

The purpose of Part 4 of this Code is to provide practical work rules as one of the means of safeguarding employees and the public from injury. It covers work rules to be followed in the installation, operation, and maintenance of electric supply and communication systems.

Supply and Communications Systems—Rules for Employees

General

1. The employer shall inform each employee working on or about communications equipment or electric-supply equipment and the associated lines, of the safety rules governing the employee's conduct while so engaged. When deemed necessary, the employer shall provide a copy of such rules.
2. Employers shall utilize positive procedures to secure compliance with these rules. Cases may arise, however, where the strict enforcement of some particular rule could seriously impede the safe progress of the work at hand; in such cases, the employee in charge of the work to be done should make such temporary modification of the rules as will accomplish the work without increasing the hazard.
3. If a difference of opinion arises with respect to the application of these rules, the decision of the employer or his authorized agent shall be final. This decision shall not result in any employee performing work in a manner which is unduly hazardous to the employee or to the employee's fellow workers.

Emergency Procedures and First Aid Rules

1. Employees shall be informed of procedures to be followed in case of emergencies and rules for first aid,

NATIONAL ELECTRICAL SAFETY CODE

including approved methods of resuscitation. Copies of such procedures and rules should be kept in conspicuous locations in vehicles and places where the number of employees and the nature of the work warrants.
2. Employees working on communications or electric-supply equipment or lines shall be regularly instructed in methods of first aid and emergency procedures, if their duties warrant such training.

Responsibility

1. A designated person shall be in charge of the operation of the equipment and lines and shall be responsible for their safe operation.
2. If more than one person is engaged in work on or about the same equipment or line, one person shall be designated as in charge of the work to be performed. Where there are separate work locations, one person may be designated at each location.

Protective Methods and Devices

Methods

1. Access to rotating or energized equipment shall be restricted to authorized personnel.
2. Diagrams, showing plainly the arrangement and location of the electric-supply equipment and lines, shall be maintained on file and readily available to authorized personnel for that portion of the system for which they are responsible.
3. Employees shall be instructed as to the character of the equipment or lines and methods to be used before any work is undertaken thereon.
4. Employees should be instructed to take additional precautions to ensure their safety when conditions create unusual hazards.

Devices and Equipment

An adequate supply of protective devices and equipment, sufficient to enable employees to meet the requirements of the work to be undertaken, and first aid equipment and materials shall be available in readily accessible and, where practical, conspicuous places.

Protective devices and equipment shall conform to the applicable standards listed in Section 3.

NOTE: The following is a list of some common protective devices and equipment, the number and kinds of which will depend upon the requirements of each case:

1. Insulating wearing apparel such as rubber gloves, rubber sleeves, and headgear
2. Insulating shields, covers, mats, and platforms
3. Insulating tools for handling or testing energized equipment or lines
4. Protective goggles
5. *Person at work* tags, portable danger signs, traffic cones, and flashers
6. Body belts and safety straps
7. Fire extinguishing equipment designed for safe use on energized parts or plainly marked that they must not be so used
8. Protective grounding materials and devices
9. Portable lighting equipment
10. First aid equipment and materials.

Inspection and Testing of Protective Devices

1. Protective devices and equipment shall be inspected or tested to ensure that they are in safe working condition.
2. Insulating gloves, sleeves, and blankets shall be inspected before use. Insulating gloves and sleeves shall be tested as frequently as their use requires.
3. Body belts, safety straps, and other personal equipment, whether furnished by employer or employee, shall be inspected to ensure that they are in safe working condition.

Warning Signs

Permanent warning signs shall be displayed in conspicuous places at all entrances to electrical-supply stations, substations, and other enclosed walk-in areas containing exposed current-carrying parts.

Identification of Supply Circuits

Means shall be provided so that identification of supply circuits can be determined before work is undertaken.

Supply Systems—Rules for Employees

Rules and Emergency Methods

The safety rules shall be carefully read and studied. Employees may be called upon at any time to show their knowledge of the rules.

Employees shall familiarize themselves with approved methods of first aid, rescue techniques, and fire extinguishment.

Safeguarding Oneself and Others

The care exercised by others should not be relied upon for protection.

1. Employees shall heed warning signs and signals and warn others who are in danger near energized equipment or lines.
2. Employees shall report promptly to the proper authority any of the following:

    a. Line or equipment defects such as abnormally sagging wires, broken insulators, broken poles, or lamp supports

    b. Accidentally energized objects such as conduits, light fixtures, or guys

    c. Other defects that may cause a dangerous condition.

3. Employees whose duties do not require them to approach or handle electric equipment and lines shall keep away from such equipment or lines and should avoid working in areas where objects and materials may be dropped by persons working overhead.
4. Employees who work on or near energized lines shall consider all of the effects of their actions, taking into account their own safety as well as the safety of other

employees on the job site, or on some other part of the affected electric system, the property of others, and the public in general.

5. Switchgear shall be de-energized prior to performing work involving removal of protective barriers unless other suitable means are provided for employee protection. The personnel safety features in switchgear shall be replaced after work is completed.

Qualification of Employees

1. Inexperienced employees working on or about energized equipment or lines shall work under the direction of an experienced and qualified person at the site.
2. Employees who do not normally work on or about electric supply lines and equipment but whose work brings them into these areas for certain tasks shall proceed with this work only when authorized by a qualified person.
3. If an employee is in doubt as to the safe performance of any assigned work, the employee shall request instructions from the employee's supervisor or person in charge.

Energized or Unknown Conditions

Electric supply equipment and lines shall be considered energized unless they are positively known to be de-energized. Before starting work, preliminary inspections or tests shall be made to determine existing conditions. Operating voltages of equipment and lines should be known before working on or near energized parts.

Ungrounded Metal Parts

All ungrounded metal parts of equipment or devices such as transformer cases and circuit breaker housings shall be considered to be energized at the highest voltage to which they are exposed, unless these parts are known by test to be free from such voltage.

Arcing Conditions

Employees should keep all parts of their bodies as far away as practical from switches, circuit breakers, or other parts at which arcing may occur during operation.

NATIONAL ELECTRICAL SAFETY CODE

Batteries

1. Enclosed areas containing storage batteries shall be adequately ventilated. Smoking, the use of open flames, and tools which may produce sparks should be avoided in such enclosed areas.
2. Employees shall use eye and skin protection when handling an electrolyte.
3. Employees shall not handle energized parts of batteries unless necessary precautions are taken to avoid shock and short circuits.

Tools and Protective Equipment

Employees shall use the personal protective equipment, the protective devices, and the special tools provided for their work. Before starting work, these devices and tools shall be carefully inspected to make sure they are in good condition.

Clothing

1. The clothing worn by an employee in the performance of his duties shall be suitable for the work to be performed and the conditions under which such work is to be performed.
2. When working in the vicinity of energized lines or equipment, the wearing of exposed metal articles such as key or watch chains, rings, wrist watches, bands, or zippers should be avoided.

Supports and Ladders

1. No employee, or any material or equipment, shall be supported or permitted to be supported on any portion of a tree, pole structure, scaffold, ladder, walkway, or other elevated structure or aerial device, etc., without it first being determined, to the extent practical, that such support is adequately strong, in good condition, and properly secured in place.
2. Portable wood ladders intended for general use shall not be painted except with a clear, nonconductive coating, nor shall they be longitudinally reinforced with metal.
3. Portable metal ladders intended for general use shall not be used when working on or near energized parts.

4. If portable ladders are made partially or entirely conductive for specialized work, necessary precautions shall be taken to ensure that their use will be restricted to the work for which they are intended.

Safety Straps

1. An employee working in an elevated position shall use a suitable safety strap or other approved means to prevent falling.
2. Safety straps or other similar devices shall be inspected by the employee to assure that they are in safe working condition.
3. Before employees trust their weight to safety straps or other devices, the employees shall determine that the snaps or fastenings are properly engaged and that the employees are secure in their body belts and safety straps.

Fire Extinguishers

In fighting fires near exposed energized parts, employees shall use fire extinguishers or materials which are suitable for the purpose. If this is not possible, all adjacent and affected equipment should first be de-energized.

Repeating Messages

Each employee receiving an oral message concerning the switching of lines and equipment shall immediately repeat it back to the sender and obtain the identity of the sender. Each employee sending such an oral message shall require it to be repeated back by the receiver and secure the latter's identity.

Machines and Moving Parts

Employees working on normally moving parts of remotely controlled equipment shall be protected against accidental starting by proper tags installed on the starting devices, and by locking or blocking where practical. Employees shall, before starting any work, satisfy themselves that these protective devices have been installed. When working near automatically or remotely operated equipment such as circuit breakers which may operate suddenly, employees shall avoid being in a position where they might be injured from such operation.

Fuses

When fuses must be installed or removed with one or both terminals energized above 1 kV, special tools insulated for the voltage or adequately rated gloves shall be used. Insulating tools or gloves should be used for voltages between 300 and 1,000. When installing expulsion type fuses, employees shall wear personal eye protection and take precautions to stand clear of the exhaust path of the fuse barrel.

Cable Reels

Cable reels shall be securely blocked so they cannot roll accidentally.

Gas-Insulated Equipment

Employees working on gas-insulated cable systems or circuit breakers shall be instructed concerning the special precautions required for possible presence of arcing by-products of SF_6.

NOTE: By-products resulting from arcing in sulfurhexaflouride (SF_6) gas-insulated systems are generally toxic and irritant. Gaseous by-products can be removed for maintenance on the compartments by purging with air or dry nitrogen. The solid residue that must be removed is mostly metallic fluoride. This fine powder absorbs moisture and produces fluorides of sulfur and hydrofluoric acid, which are toxic and corrosive.

Operating Routines

Duties of a Designated Person

A designated person shall

1. Keep informed of operating conditions affecting the safe and reliable operation of the system
2. Maintain a suitable record showing operating changes in such conditions.

Duties of a First-Line Supervisor

A first-line supervisor shall

1. Adopt such precautions as are within the first-line supervisor's power to prevent accidents and to see that

the safety rules and operating procedures are observed by the employees under the direction of the first line supervisor.
2. Make all the necessary records and report to the designated person when required.
3. As far as possible, prevent unauthorized persons from approaching places where work is being done.
4. Prohibit the use of any tools or devices unsuited to the work at hand or which have not been tested or inspected as required by these rules.

Guides

Persons accompanying uninstructed employees or visitors near electric equipment or lines shall be qualified to safeguard the people in their care and see that the safety rules are observed.

Authorization

Specific work Authorization from the designated person shall be secured before work is begun on or about station equipment, transmission, or interconnected feeder circuits and where circuits are to be de-energized at stations. The designated person shall be notified when such work ceases. EXCEPTION: In an emergency, to protect life or property or when communication with the designated person is difficult because of storms or other causes, any qualified employee may make repairs on or about the equipment or lines covered by this rule without special authorization if the qualified employee can clear the trouble promptly with available help in compliance with the remaining rules. The designated person thereafter be notified as soon as possible of the action taken.

Operations at stations In the absence of specific operating schedules, employees shall secure authorization from the designated person before opening and closing supply circuits or starting and stopping equipment affecting system operation at stations. EXCEPTION: In an emergency, to protect life or property, any qualified employee may open circuits and stop moving equipment without special authorization if, in the judgement of the qualified employee, this action will promote safety, but the designated person shall be notified as soon as possible of such action, with reasons therefore.

De-energizing sections of circuits sectionalizing devices
Employees shall obtain authorization from the designated person before de-energizing sections of circuits. EXCEPTION: Sections of distribution circuits are excepted if the designated person is notified as soon as possible after the action is taken.

Protecting Employees by Switches and Disconnectors

When equipment or lines are to be disconnected from any source of electric energy for the protection of employees, the switches, circuit breakers, or other devices designated and designed for operation under the load involved at sectionalizing points shall be opened or disconnected first.

Re-energizing After Work

Instructions to re-energize equipment or lines which have been de-energized by permission of the designated person shall not be issued by the designated person until all employees who requested the lines to be de-energized have reported clear. Employees who have requested equipment or lines de-energized for other employees or crews shall not request that equipment or lines be re-energized until all of the other employees or crews have reported clear. The same procedure shall be followed when more than one location is involved.

Tagging Electric-Supply Circuits

Equipment or circuits that are to be treated as de-energized shall have suitable tags attached to all points where such equipment or circuits can be energized. Controls that are to be de-activated during the course of work on energized or de-energized equipment or circuits also shall be tagged. The tags shall be placed to identify plainly the equipment or circuits on which work is being performed.

Maintaining Service

1. *Closing tagged circuits which have opened automatically.* When controls upon which tags have been placed open automatically, they shall be left open until reclosing has been authorized.
2. *Closing circuits opened automatically.* When circuits open automatically, local operating rules shall

determine in what manner and how many times they may be closed with safety.

3. *Unintentional grounds on delta circuits.* Unintentional grounds on delta circuits shall be removed as soon as practical.

Area Protection

Vehicular and pedestrian traffic

a. Before engaging in work that may endanger the public, warning signs or traffic control devices, or both, shall be placed conspicuously to approching traffic. Where further protection is needed, suitable barrier guards shall be erected. Where the nature of work and traffic requires it, a person shall be stationed to warn traffic while the hazard exists.

b. In case openings or obstructions in the street, sidewalk, walkways, or on private property are being worked on or left unattended during the day, danger signals, such as warning signs and flags, shall be effectively displayed. Under these same conditions at night, warning lights shall be prominently displayed and excavations shall be enclosed with protective barricades.

Employees

a. If the work exposes energized or moving parts that are normally protected, danger signs shall be displayed and suitable guards erected to warn other personnel in the area.

b. When working in one section where there is a multiplicity of such sections, such as one panel of a switchboard, one compartment of several, or one portion of a substation, personnel shall mark the work area conspicuously and place barriers to prevent accidental contact with energized parts in that section or adjacent sections.

Crossed or fallen wires An employee finding any crossed or fallen wires which are or may create a hazard shall remain on guard or initiate appropriate action. If the employee can observe the rules for handling energized parts by the use of insulating equipment, that employee may correct the condition at once; otherwise it shall be necessary to secure authorization first for so doing.

NATIONAL ELECTRICAL SAFETY CODE

Handling Energized Equipment or Lines

General Requirements

Work on energized lines and equipment When working on energized lines and equipment, one of the following safeguards shall be applied:

a. Insulate employee from energized parts
b. Isolate or insulate the employee from ground and grounded structures and potentials other than the one being worked on.

Covered (noninsulated) wire Employees should not place dependence for their safety on the covering of wires. All precautions in this section for handling energized parts shall be observed in handling covered wires.

Exposure to higher voltages All employees working on or about equipment or lines exposed to voltages higher than those guarded against by the safety appliances provided shall assure themselves that the equipment or lines on which they are working are free from dangerous leakage or induction or have been effectively grounded.

Cutting into insulating coverings of energized conductors When the insulating covering on energized wires or cable must be cut into, the employee shall use suitable tools. While doing such work, suitable eye protection and insulating gloves with protectors shall be worn. Employees shall exercise extreme care to prevent short-circuiting conductors when cutting into the insulation.

Metal tapes or ropes Metal measuring tapes, and tapes or ropes containing metal threads or strands, shall not be used closer to exposed energized parts than the distance specified in this Code. Also, care should be taken when extending metallic ropes or tapes parallel to and in the proximity of high volume lines because of the effect of induced voltages.

Work equipment or material extending into energized areas Equipment or material of a noninsulating substance which is not bonded to an effective ground and extends into an energized area, and could approach energized equipment closer than the distance specified in Rule 422B, shall be treated as though it is energized at the same voltage as the line or equipment to which it is exposed.

Table 5.1 Illumination Levels

Location	(footcandles)	(lux)
Central Station		
Air conditioning equipment, air preheater and fan floor, ash sluicing	5	55
Auxiliaries, battery areas, boiler feed pumps, tanks, compressors, gage area	10	110
Boiler platforms	5	55
Burner platforms	10	110
Cable room, circulator, or pump bay	5	55
Chemical laboratory	25	270
Coal conveyor, crusher, feeder, scale areas, pulverizer, fan area, transfer tower	5	55
Condensers, deaerator floor, evaporator floor, heater floors	5	55
Control rooms		
Vertical face of switchboards		
Simples or section of duplex operator:		
Type A—Large centralized control room 66 inches above floor	25	270
Type B—Ordinary control room 66 inches above floor	15	160
Section of duplex facing away from operator	15	160
Bench boards (horizontal level)	25	270
Area inside duplex switchboards	5	55
Rear of all switchboard panels (vertical)	5	55
Dispatch boards		
Horizontal plane (desk level)	25	270
Vertical face of board (48 inches above floor, facing operator):		
System load dispatch room	25	270
Secondary dispatch room	15	160
Hydrogen and carbon dioxide manifold area	10	110
Precipitators	5	55
Screen house	10	110
Soot or slag blower platform	5	55
Steam headers and throttles	5	55

Table 5.1 Illumination Levels *(Continued)*

Location	(footcandles)	(lux)
Switchgear, power	10	110
Telephone equipment room	10	110
Tunnels or galleries, piping	5	55
Turbine bay sub-basement	10	110
Turbine room	15	160
Visitors' gallery	10	110
Water treating area	10	110
Central Station (Exterior)		
Catwalks	2	22
Cinder dumps	0.2	2.2
Coal storage area	0.2	2.2
Coal unloading		
Dock (loading or unloading zone)	5	55
Barge storage area	0.5	5.5
Car dumper	0.5	5.5
Tipple	5	55
Conveyors	2	22
Entrances		
Generating or service building		
Main	10	110
Secondary	2	22
Gate house		
Pedestrian entrance	10	110
Conveyor entrance	5	55
Fence	0.2	2.2
Fuel-oil delivery headers	5	55
Oil storage tanks	1	11
Open yard	0.2	2.2
Platforms—boiler, turbine deck	5	55
Roadway		
Between or along buildings	1	11
Not bordered by buildings	0.5	5.5
Substation		
General horizontal	2	22
Specific vertical (on disconnects)	2	22

Clearance from Live Parts

No employee shall approach or take any conductive object without a suitable insulating handle within the distances of any exposed energized part listed in Tables 2 and 3 unless the employee is insulated from the energized part, the energized part is insulated from the employee, or the employee is insulated from all conducting surfaces other than the one upon which the employee is working. Gloves rated for the voltage involved shall be considered effective insulation of the employee from the energized part.

Requirement for Assisting Employee

In inclement weather or at night, no employee shall work alone outdoors on or dangerously near energized conductors or parts

Table 5.2 AC Minimum Clearance from Live Parts

Nominal voltage in kilovolts phase to phase	Distance phase to employee (ft)	(m)
1 to 34.5	2	0.60
46	2½	0.75
69	3	0.90
115	3	0.90
138	3½	1.07
161	3½	1.07
230	5	1.50

Table 5.3 DC Minimum Clearance from Live Parts

Maximum Voltage Conductor to Ground (kilovolts)	Distance (ft)	(m)
250	3½	1.07
400	6	1.80
500	8½	2.60
750	16	4.9

NOTE: These distances are based on the highest transient over-voltage and employee will be exposed to on any system with live-line tools as the insulating material.

of more than 750 V between conductors. EXCEPTION: This shall not preclude a qualified employee, working alone, from cutting trouble in the clear, switching, replacing fuses, or similar work if such work can be performed safely.

When to De-energize Parts

An employee shall not approach, or willingly permit others to approach, any exposed ungrounded part normally energized, unless the supply equipment or lines are de-energized.

Opening and Closing Switches

Manual switches and disconnectors shall always be closed by a continuous motion. Care should be exercised in opening switches to avoid serious arcing.

Working Position

Employees should avoid working on equipment or lines in any position from which a shock or slip will tend to bring the body toward exposed parts at a potential different than the employee's body. Work should, therefore, generally be done from below, rather than from above.

Making Connections

In connecting de-energized equipment or lines to an energized circuit by means of a conducting wire or device, employees should first attach the wire to the de-energized part. When disconnecting, the source end should be removed first. Loose conductors should be kept away from exposed energized parts.

Current Transformer Secondaries

The Secondary of a current transformer shall not be opened while energized. If the entire circuit cannot be properly de-energized before working on an instrument, a relay, or other section of a current transformer secondary circuit, the employee shall bridge the circuit with jumpers so that the current transformer secondary will not be opened.

Capacitors

Before employees work on capacitors, the capacitors shall be disconnected from the energizing source, short-circuited, and grounded. Any line to which capacitors are connected shall be short-circuited and grounded before it is considered de-energized. Since capacitor units may be connected in series-parallel, each unit shall be shorted between all insulated terminals and the capacitor tank before handling. Where the tanks of capacitors are on ungrounded racks, the racks shall also be grounded. The internal resistor shall not be depended upon to discharge capacitors.

De-energizing Equipment or Lines to Protect Employees

Application of Rule

1. When employees must depend on others to operate switches or otherwise de-energize circuits on which they are to work, or must secure special authorization before they operate such switches themselves, the precautionary measures that follow shall be taken in the order given before work is begun.
2. If the employee under whose direction a section of a circuit is disconnected is in sole charge of the section and of the means of disconnection, those portions of the measures that follow which pertain to dealing with the designated person may be omitted.
3. Records shall be kept on all contractural utility interactive systems on any electric supply lines. When these lines are de-energized according to Rule 423C, the utility interactive system shall be visibly disconnected from the lines.

Employee's Request

The employee in charge of the work must apply to the designated person to have the particular section of equipment or lines de-energized, identifying it by position, letter color, number, or other means.

Opening Disconnectors and Tagging

The designated person shall direct the opening of all switches and disconnectors through which electric energy may be supplied to the particular section of equipment and lines to be de-energized, and shall direct that such switches and disconnectors be rendered inoperable and tagged, plainly indicating that persons are at work. If switches that are controlled automatically or remotely, or both, can be rendered inoperable, they shall be tagged at the switch location. If it is impractical to render such switches and disconnectors inoperable, then these remotely controlled switches shall also be tagged at all points of control. A record shall be made when placing the tag giving the time of disconnection, the name of the person making the disconnection, the name of the employee who requested the disconnection, and the name or title, or both, of the designated person.

Employee's Protective Grounds

When all the switches and disconnectors designated have been opened, rendered inoperable where practical, and tagged, and the employee has been given permission to work by the designated person, the employee in charge should immediately proceed to make the employee's own protective grounds or verify that adequate grounds have been applied on the disconnected lines or equipment. During the testing for potential and/or application of grounds, distances not less than the Code's, as applicable, shall be maintained.

Grounds shall be placed at each side of the work location and as close as practical to the work location, or a single ground point shall be placed at the work location. If work is to be performed at more than one location on a line section, the line section shall be grounded and short circuited at one location in the line section, and the conductor to be worked on shall be grounded at each work location.

The Code distances shall be maintained from ungrounded conductors at the work location. Where the making of a ground is impractical, or the conditions resulting therefrom are more hazardous than working on the lines or equipment without grounding, the ground may be omitted by special permission of the designated person.

Proceeding with Work

1. After the equipment or lines have been de-energized and grounded, the employee in charge and those under the direction of the employee in charge may proceed with work on the de-energized parts. Equipment may be re-energized for testing purposes only under the

supervision of the employee in charge and subject to authorization by the designated person.

2. Each additional employee in charge desiring the same equipment or lines to be de-energized for the protection of that person, or the persons under direction, shall follow these procedures to secure similar protection.

Reporting Clear—Transferring Responsibility

1. The employee in charge, upon completion of the work and after assuring that all persons assigned to this employee in charge are in the clear, shall remove protective grounds and shall report to the designated person that all tags protecting that person may be removed.
2. The employee in charge who received the permission to work may, if specifically permitted by the designated person, transfer the permission to work and the responsibility for persons by personally informing the affected persons of the transfer.

Removal of Tags

1. The designated person shall then direct the removal of tags and the removal shall be reported back to the designated persons by the persons removing them. Upon the removal of any tag, there shall be added to the record containing the name of the designated person or title, or both, and, the person who requested the tag, the name of the person requesting removal, the time of removal, and the name of the person removing the tag.
2. The name of the person requesting removal shall be the same as the name of the person requesting placement, unless responsibility has been transferred according to Rule 423F.

Restoring Service

Only after all protective grounds have been removed from the circuit or equipment and after all protective tags have been removed at a specific location, may the designated person direct the closing of disconnectors and switches at that location.

NATIONAL ELECTRICAL SAFETY CODE

Protective Grounds

Installing Grounds

When placing temporary protective grounds on a normally energized circuit, the following precautionary measures shall be observed.

Size of grounds The grounding device shall be of such size as to carry the induced current and maximum fault current that could flow at the point of grounding for the time necessary to clear the line.

Ground connections The employee making a protective ground on equipment or lines shall first connect one end of the grounding device to an effective ground connection.

Test of circuit The de-energized conductors and equipment which are to be grounded shall next be tested for voltage except where previously installed grounds are clearly in evidence. The employee shall keep every part of the body at the required distance by using insulating handles of proper length or other suitable devices.

Completing grounds If the test shows no voltage, or the local operating rules so direct, the free end of the grounding device shall next be brought into contact with the de-energized part using insulating handles or other suitable devices and securely clamped or otherwise secured thereto. Where bundled conductor lines are being grounded, grounding of each subconductor should be made. Only then may the employee come within the distances from the de-energized parts or proceed to work upon the parts as upon a grounded part.

NOTE: In stations, switches may be employed to connect the equipment or lines being grounded to the actual ground connection.

Removing Grounds

The employee shall first remove the grounding device from the de-energized parts using insulating handles or other suitable devices.

Overhead Lines

Employees working on or with overhead lines shall observe the following rules in addition to applicable rules contained elsewhere in this Code.

Checking Structures Before Climbing

1. Before climbing poles, ladders, scaffolds, or other elevated structures, employees shall determine, to the extent practical, that the structures are capable of sustaining the additional or unbalanced stresses to which they will be subjected.
2. Where there are indications that poles and structures may be unsafe for climbing, they shall not be climbed until made safe by guying, bracing, or other means.

Installing and Removing Wires or Cables

1. Precautions shall be taken to prevent wires or cables being installed or removed from contacting energized wires or equipment. Wires or cables which are not bonded to an effective ground and are being installed or removed in the vicinity of energized conductors shall be considered as being energized.
2. Sag of wires or cable being installed or removed shall be controlled to prevent danger to pedestrian and vehicular traffic.
3. Before installing or removing wire or cable, strains to which poles and structures will be subjected shall be considered and necessary action taken to prevent failure of support in structures.
4. Employees should avoid contact with moving winch lines, especially near sheaves, blocks, and take-up drums.

Setting, Moving, or Removing Poles in or Near Energized Lines

1. When setting, moving, or removing poles in or near energized lines, precautions shall be taken to avoid direct contact of the pole with the energized conductors. Employees shall wear suitable insulating gloves or use other suitable means where voltages may exceed rating of gloves in handling poles where conductors energized at potentials above 750 V can be contacted. Employees performing such work shall not contact the pole with uninsulated parts of their bodies.
2. Contact with trucks, derricks, or other equipment which are not bonded to an effective ground and is being used

to set, move, or remove poles in or near energized lines shall be avoided by employees standing on the ground or in contact with grounded objects unless employees are wearing suitable protective equipment.

Underground Lines

Employees working on or with underground lines shall observe the following rules in addition to applicable rules contained elsewhere in this Code.

Guarding Manhole and Street Openings

When covers of manholes, handholes, or vaults are removed, the opening shall be promptly protected with a barrier, temporary cover, or other suitable guard.

Testing for Gas in Manholes and Ventilated Vaults

1. The atmosphere shall be tested for combustible or flammable gas(es) before entry.
2. Where combustible or flammable gas(es) are detected, the work area shall be ventilated and made safe before entry.
3. Unless forced continuous ventilation is provided, a test shall also be made for oxygen deficiency.
4. Provision shall be made for an adequate continuous supply of air.
   NOTE: The term adequate includes evaluation of both the quantity and quality of the air.

Attendant on Surface

While personnel are in a manhole, an employee shall be available on the surface in the immediate vicinity primarily to rendered assistance from the surface. This shall not preclude the employee on the surface from entering the manhole to provide short-term assistance. EXCEPTION: This shall not preclude a qualified employee, working alone, from entering a manhole where energized cables or equipment are in service, for the purpose of inspection, housekeeping, taking readings, or similar work if such work can be performed safely.

Flames

1. Employees shall not smoke in manholes.

2. Where open flames must be used in manholes or vaults, extra precautions shall be taken to ensure adequate ventilation.
3. Before using open flames in an excavation in areas where combustible gases or liquids may be present, such as near gasoline service stations, the atmosphere of the excavation shall be tested and found safe or cleared of the combustible gases or liquids.

Excavation

1. Cables and other buried utilities in the immediate vicinity shall be located, to the extent practical, prior to excavating.
2. Hand tools used for excavating near energized supply cables should be equipped with handles made of nonconductive material.
3. Mechanized equipment should not be used to excavate in close proximity to cables and other buried utilities.
4. If a gas or fuel line is broken or damaged, employees shall

   a. Leave the excavation open

   b. Extinguish flames which could ignite the escaping gas or fuel

   c. Notify the proper authority

   d. Keep the public away until the condition is under control.

Identification

1. When underground facilities are exposed (electric, gas, water, telephone, etc.), they should be identified and shall be protected as necessary to avoid damage.
2. Where multiple cables exist in an excavation, cables other than the one being worked on shall be protected as necessary.
3. Before cutting into a cable or opening a splice, the cable should be identified and verified to be the proper cable.
4. A cable to be worked on as de-energized which cannot be positively identified or determined to be de-energized

shall be pierced or severed with an approved tool at the work location.

5. Before cutting into an energized cable, the operating voltage shall be determined and appropriate precautions taken for handling conductors at that voltage.

Operation of Power Driven Equipment

Employees should avoid being in manholes where power driven rodding equipment is in operation.

Live-Line Bare-Hand Work

All employees using live-line bare-hand work practices shall observe the following rules in addition to applicable rules contained elsewhere in this Code.

Training

Employees shall be trained in live-line bare-hand work methods before being permitted to use this technique on energized lines.

Equipment

1. Insulated aerial devices used in bare-hand work shall be tested before the work is started to assure the integrity of the insulation.
2. Insulated aerial devices and other equipment used in this work shall be maintained in a clean condition.
3. Tools and equipment shall not be used in a manner that will reduce the overall insulating strength of the insulated aerial device.

Bonding and Shielding

1. A conductive bucket liner or other suitable conducting device shall be provided for bonding the insulated aerial device to the energized line or equipment.
2. The employee shall be bonded to the insulated aerial device by use of conducting shoes, leg clips, or other suitable means.
3. Adequate electrostatic shielding shall be provided and used where necessary.

4. Before the employee contacts the energized part to be worked on, the aerial device shall be bonded to the energized conductor by means of a positive connection.

General Precautions

Rules and Emergency Methods

The safety rules shall be carefully read and studied. Employees may be called upon at any time to show their knowledge of the rules.

Employees shall familiarize themselves with approved methods of first aid, rescue techniques, and fire extinguishment.

Safeguarding Oneself and Others

The care exercised by others should not be relied upon for protection.

1. Employees shall heed warning signs and signals and warn others who are in danger near energized equipment or lines.
2. Employees shall report promptly to the proper authority any of the following:

    a. Line or equipment defects such as abnormally sagging wires, broken insulators, broken poles, or lamp supports

    b. Accidentally energized objects such as conduit, light fixtures, or guys

    c. Other defects that may cause a dangerous condition.

3. Employees whose duties do not require them to approach or handle electric equipment and lines shall keep away from such equipment or lines and should avoid working in areas where objects and materials may be dropped by persons working overhead.
4. Employees who work near energized supply lines shall consider all the effects of their actions, taking into account their own safety as well as the safety of other employees on the job site, the property of others, and the public in general.

NATIONAL ELECTRICAL SAFETY CODE

5. No employee shall approach or take any conductive object within the distances of any exposed energized part listed in Table 4.
6. Care should be exercised when extending metal ropes, tapes, or wires parallel to and in the proximity of energized high voltage lines because of induced voltages. When it is necessary to measure clearances from energized objects, only devices approved for the purpose shall be used.

Qualifications of Employees

1. Inexperienced employees working in the vicinity of energized electric-supply equipment or lines shall work under the direction of an experienced and qualified person at the site.
2. Employees who do not normally work in the vicinity of electric-supply lines and equipment but whose work brings them into these areas for certain tasks, shall proceed with this work only when authorized by a qualified person.
3. If an employee is in doubt as to the safe performance of any work assigned, instructions shall be requested from the employee's supervisor or person in charge.

Table 5.4 Approach Distances to Exposed Energized Overhead Power Lines and Parts

Voltage range (phase-to-phase, RMS)	Approach distance (inches)
300 V and less	①
Over 300 V, not over 750 V	12
Over 750 V, not over 2 kV	18
Over 2 kV, not over 15 kV	24
Over 15 kV, not over 37 kV	36
Over 37 kV, not over 87.5 kV	42
Over 87.5 kV, not over 121 kV	48
Over 121 kV, not over 140 kV	54

① Avoid contact.

Energized or Unknown Conditions

Electric-supply equipment and lines shall be considered to be energized unless they are known to be de-energized. Operating voltages of such equipment and lines should be known before working near energized parts.

Ungrounded Metal Parts

All ungrounded metal parts of equipment, such as transformer cases and circuit breaker housings, shall be considered energized at the highest voltage to which they are exposed, unless these parts are known by test to be free from such voltage.

Arcing Conditions

Employees should keep all parts of their bodies as far away as practical from brushes, commutators, switches, circuit breakers, or other parts at which arcing may occur during operation or handling.

Batteries

1. Enclosed areas containing storage batteries shall be adequately ventilated. Smoking, the use of open flames, and tools which may produce sparks should be avoided in such enclosed areas.
2. Employees shall use eye and skin protection when handling an electrolyte.
3. Employees shall not handle energized parts of batteries unless necessary precautions are taken to avoid shock and short circuits.

Tools and Protective Equipment

Employees shall use the personal protective equipment, the protective devices, and the special tools provided for their work. Before starting work, these devices and tools shall be carefully inspected to make sure that they are in good condition.

Clothing

The clothing worn by an employee in the performance of the employee's duties shall be suitable for the work to be performed and the conditions under which such work is to be performed.

Supports and Ladders

1. No employee, or any material or equipment, shall be supported or permitted to be supported on any portion of a tree, pole structure, scaffold, ladder, walkway, or other elevated structure or aerial device, etc., without it first being determined, to the extent practical, that such support is adequately strong, in good condition, and properly secured in place.
2. Portable wood ladders intended for general use shall not be painted except with a clear nonconductive coating, nor shall they be longitudinally reinforced with metal.
3. Portable metal ladders shall not be used when working near energized parts of electric-supply systems.
4. If portable ladders are made partially or entirely conductive for specialized work, necessary precautions shall be taken to ensure that their use will be restricted to the work for which they are intended.

Safety Straps

1. An employee working in an elevated position shall use a suitable safety strap or other approved means to prevent falling.
2. Safety straps or other devices shall be inspected by the employee to assure that they are in safe working condition.
3. Before employees trust their weight to safety straps or other devices, the employees shall determine that the snaps or fastenings are properly engaged and that the employees are secured in their body belts and safety straps.

Fire Extinguishers

In fighting fires near exposed energized parts of electric supply system, employees shall use fire extinguishers or materials which are suitable for the purpose. If this is not possible, all adjacent and affected equipment should first be de-energized.

Machines and Moving Parts

Employees working on normally moving parts of remotely controlled equipment shall be protected against accidental starting by proper tags installed on the starting devices, and by locking or blocking where practical. Employees shall, before starting any work, satisfy themselves that these protective devices have been installed. When working near automatically or remotely operated equipment such as circuit breakers which may operate suddenly, employees shall avoid being in a position where they might be injured from such operation.

Fuses

When fuses must be installed or removed with one or both terminals energized, special tools insulated for the voltage shall be used.

Cable Reels

Cable reels shall be securely blocked so they cannot roll accidentally.

Operating Routines

Duties of a First-Line Supervisor

A first-line supervisor shall

1. Adopt such precautions as are within the first-line supervisor's power to prevent accidents and see that the safety rules and operating procedures are observed by the employees under the direction of the first-line supervisor.
2. As far as possible, prevent unauthorized persons from approaching places where work is being done.
3. Prohibit the use of any tools or devices unsuited to the work at hand or which have not been tested or inspected as required by these rules.

Area Protection

Vehicular and pedestrian traffic

    a. Before engaging in work that may endanger the public, warning signs or traffic control devices, or both, shall be placed conspicuously to approaching traffic. Where further protection is needed, suitable barrier guards shall be erected. Where the nature of work and traffic requires it, a person shall be stationed to warn traffic while the hazard exists.

b. In case openings or obstructions in the street, sidewalk, walkway, or on private property are being worked on or left unattended during the day, danger signals, such as warning signs and flags, shall be effectively displayed.

Under these same conditions at night, warning lights shall be prominently displayed, and excavations shall be enclosed with protective barricades.

Employees If the work exposes energized or moving parts that are normally protected, danger signs shall be displayed and suitable guards erected to warn other personnel in the areas.

Crossed or fallen wires An employee finding any crossed or fallen wires that are creating, or may create, a hazard shall remain on guard or adopt other adequate means to prevent accidents, and shall have the proper authority notified.

Overhead Lines

All employees working on or with overhead lines shall observe the following rules in addition to applicable rules contained elsewhere in this Code.

Checking Structures Before Climbing

1. Before climbing poles, ladders, scaffolds, or other elevated structures, employees shall determine, to the extent practical, that the structures are capable of sustaining the additional or unbalanced stresses to which they will be subjected.
2. Where there are indications that poles and structures may be unsafe for climbing, they shall not be climbed until made safe by guying, bracing, or other means.
3. Employees shall not climb poles where electric-supply wires or equipment are hanging below their proper level.

Position on Poles

When working on jointly used poles or structures, employees shall not climb or work above the level of the lowest electric-supply conductor exclusive of vertical runs and street light wiring. EXCEPTION: This rule does not apply where communications facilities are attached above electric-supply conductors if a rigid fixed barrier has been installed between the supply and communications facilities.

Installing and Removing Wires and Cables

1. Precautions shall be taken to prevent wires or cables being installed or removed from contacting energized wires or equipment. Wires or cables which are not bonded to an effective ground and are being installed or removed in the vicinity of energized conductors shall be considered as being energized.
2. Sag of wire or cables being installed or removed shall be controlled to prevent danger to pedestrian and vehicular traffic.
3. Before installing or removing wires or cables, the strains to which poles and structures will be subjected shall be considered and necessary action taken to prevent failure of supporting structures.
4. Employees should avoid contact with moving winch lines, especially near sheaves, blocks, and take-up drums.
5. Every employee working on or about equipment or lines exposed to voltages higher than those guarded against by the safety appliances provided shall take steps to be assured that the equipment or lines on which the employee is working are free from dangerous leakage or induction or have been effectively grounded.

Setting, Moving, or Removing Poles in or Near Energized Electric-Supply Lines

1. When setting, moving, or removing poles in or near energized lines, precautions shall be taken to avoid direct contact of the pole with the energized conductors. Employees shall wear suitable insulating gloves or use other suitable means where voltage may exceed rating or gloves in handling poles where conductors energized at potentials above 750 V can be contacted. Employees performing such work shall not contact the pole with uninsulated parts of their body.

2. Contact with trucks, derricks, or other equipment which are not bonded to an effective ground and are being used to set, move, or remove poles in or near energized lines shall be avoided by employees standing on the ground, or in contact with grounded objects, unless the employee is wearing suitable protective equipment.

Underground Lines

Employees working on or with underground lines shall observe the following rules in addition to applicable rules contained elsewhere in this Code.

Guarding Manhole and Street Openings

When covers of manholes, handholes, or vaults are removed, the opening shall be promptly protected with a barrier, temporary cover, or other suitable guard.

Testing for Gas in Manholes and Unventilated Vaults

1. The atmosphere shall be tested for combustible or flammable gas(es) before entry.
2. Where combustible or flammable gas(es) are detected, the work area shall be ventilated and made safe before entry.
3. Unless forced continuous ventilation is provided, a test shall also be made for oxygen deficiency.
4. Provision shall be made for an adequate continuous supply of air.
   NOTE: The term adequate includes evaluation of both the quantity and the quality of the air.

Attendant on Surface

While personnel are in a joint-use manhole, an employee shall be available on the surface in the immediate vicinity to render assistance as may be required.

Flames

1. Employees shall not smoke in manholes.
2. Where open flames must be used in manholes or vaults, extra precautions shall be taken to ensure adequate ventilation.

3. Before using open flames in an excavation in areas where combustible gases or liquids may be present, such as near gasoline service stations, the atmosphere of the excavation shall be tested and found safe or cleared of the combustible gases or liquids.

Excavation

1. Cables and other buried utilites in the immediate vicinity shall be located to the extent practical prior to excavating.
2. Hand tools used for excavating near energized supply cables shall be equipped with handles made of nonconductive material.
3. Mechanized equipment should not be used to excavate in close proximity to cables and other buried utilities.
4. If a gas or fuel line is broken or damaged, employees shall

    a. Leave the excavation open

    b. Extinguish flames which could ignite the escaping gas or fuel

    c. Notify the proper authority

    d. Keep the public away until the condition is under control.

Identification

1. When underground facilities are exposed (electric, gas, water, telephone, etc.), they should be identified and shall be protected as necessary to avoid damage.
2. Where multiple cables exist in an excavation, cables other than the one being worked on shall be protected as necessary.
3. When multiple cables exist in an excavation, the cable to be worked on shall be identified by electrical means unless its identity is obvious by reason of distinctive appearance.
4. Before cutting into a cable or opening a splice in a joint use manhole, the cable shall be identified and verified to be the proper cable.

NATIONAL ELECTRICAL SAFETY CODE

Maintaining Sheath Continuity

When working on buried cable or on cable in manholes, sheath continuity shall be maintained by bonding across the opening or by equivalent means.

Operation of Power-Driven Equipment

Employees should avoid being in manholes where power-driven rodding equipment is in operation.

6

Federal Regulations

In 1970, the Congress of the United States and the President placed into law the Williams-Steiger Occupational Safety and Health (OSHA) Act to protect employees from hazards of the workplace. In the Act, Congress declared it to be its purpose and policy "... to assure so far as possible every working man and woman in the Nation safe and healthful working conditions and to preserve our human resources" by, among other actions and programs, "encouraging the States to assume the fullest responsibility for administration and enforcement of their occupational and health laws...." The Act also provides that any State which desires to assume responsibility for the development and enforcement of occupational safety and health standards to issues covered by corresponding standards placed into law by the Act shall submit a plan for doing so to the Assistant Secretary of Labor (for Title 29 of the Act relating to the Department of Labor).

The Code of Federal Regulations (CFR) is a codification of the general and permanent rules published in the *Federal Register* by the Executive Departments and agencies of the federal government. The Code is divided into 50 titles (Title 29 relates to safety in the workplace and administered by the Department of Labor) which represent broad areas subject to federal regulation. Each title is divided into chapters which usually bear the name of the issuing agency. Each chapter is further subdivided into parts covering specific regulatory areas. Each volume of the Code is revised at least each calendar year and issued on a quarterly basis, for example, as of July 1 for Title 29.

Title 29 is composed of seven volumes. The parts in these volumes are arranged in the following order: Parts 900-1899, Parts 1900-1910, Parts 1911-1919, and Parts 1920 to end. The contents of these volumes represent all current regulations codified under Title 29 as of July 1, 1985.

Chapter VIII of Title 29, Subtitle B, relates to "Occupational Safety and Health Administration, Department of Labor" (Parts 1900 to 1910). The Parts are assigned subjects as follows:

Part	Subject
1900	(Reserved)
1901	Procedures for State agreements
1902	State plans for the development and enforcement of State standards
1903	Inspections, citations, and proposed penalties.
1904	Recording and reporting occupational injuries and illnesses.
1905	Rules of practice for variances, limitations, variations, tolerances, and exemptions under the Act.
1906	Administrative witnesses and documents in private litigation (Reserved)
1907	Accreditation of testing laboratories
1908	Consultation agreements
1910	Occupational safety and health standards.

Parts 1900 through 1908 are subjects not requiring many pages (75) of Title 29 of the CFR, and there is no Part 1908 (no explanation, either). Part 1910 takes up 863 pages of the Title. Subpart S of Title 29 covers the electrical aspects of the Act. The decimal system of the Parts for the electrical standards starts at 1910.301 and ends at 1910.399, some 35 pages in length.

The complete electrical OSHA regulations are presented to familiarize users of electrical equipment or systems with the wording and content.

FEDERAL REGULATIONS

GENERAL

1910.301 Introduction

This subpart addresses electrical safety requirements that are necessary for the practical safeguarding of employees in their workplaces and is divided into four major divisions as follows:

(a) Design safety standards for electrical systems.

(b) Safety-related work practices.

(c) Safety-related maintenance requirements.

(d) Safety requirements for special equipment.

(e) Definitions.

DESIGN SAFETY STANDARDS FOR ELECTRICAL SYSTEMS

1910.302 Electric utilization systems.

(a) Scope

(1) Covered. The provisions of 1910.302 through 1910.308 of this subpart cover electrical installations and utilization equipment installed or used within or on buildings, structures, and other premises, including the following:

Yards

Carnivals

Parking and other lots

Mobile homes

Recreational vehicles

Industrial substations

Conductors that connect the installations to a supply of electricity, and

Other outside conductors on the premises.

(2) **Not covered.** The provisions of 1910.302 through 1910.308 of this subpart do not cover the following:

(i) Installations in ships, watercraft, railway rolling stock, aircraft, or automotive vehicles other than mobile homes and recreational vehicles.

(ii) Installations underground in mines.

(iii) Installations of railways for generation, transformation, transmission, or distribution of power used exclusively for operation of rolling stock or installations used exclusively for signaling and communication purposes.

(iv) Installations of communication equipment under the exclusive control of communication utilities, located outdoors or in building spaces used exclusively for such installations.

(v) Installations under the exclusive control of electric utilities for the purpose of communication or metering; or for the generation, control, transformation, transmission, and distribution of electric energy located in buildings used exclusively by utilities for such purposes or located outdoors on property owned or leased by the utility or on public highways, streets, roads, etc., or outdoors by established rights on private property.

(b) **Extent of application**

(1) The requirements contained in the sections listed below shall apply to all electrical installations and utilization equipment, regardless of when they were designed or installed.

Sections:

1910.303	Examination, installation, and use of equipment.
1910.304	Protection of conductors and equipment.
1910.305	Flexible cords and cables, uses.
1910.307	Hazardous (classified) locations.

(2) Every electric utilization system and all utilization equipment installed after March 15, 1972, and every major replacement, modification, repair, or rehabilitation, after March 15, 1972, of any part of any electric utilization system or utilization equipment installed before March 15, 1972, shall comply with the provisions of 1910.302 through 1910.308.

Note: "Major replacements, modifications, repairs, or rehabilitations" include work similar to that involved when a new building or facility is built, a new wing is added, or an entire floor is renovated.

(3) The following provisions apply to electric utilization systems and utilization equipment installed after April 16, 1981:

1910.303(h)(4)(i) and (ii)	Entrance and access to workspace (over 600 volts)
1910.304(e)(1)(vi)(b)	Circuit breakers operated vertically
1910.304(e)(1)(vi)(c)	Circuit breakers used as switches
1910.304	Grounding of systems of 1000 volts or more supplying portable or mobile equipment
1910.305	Switching series capacitors over 600 volts.
1910.306	Warning signs for elevators and escalators; Electrically controlled irrigation machines; Ground-fault circuit interrupters for fountains
1910.308	Physical protection of conductors over 600 volts; Marking of Class 2 and Class 3 power supplies; Fire protective signaling circuits.

1910.303 General requirements.

(a) **Approval.** The conductors and equipment required or permitted by this subpart shall be acceptable only if approved.

(b) **Examination, installation, and use of equipment.**

(1) **Examination.** Electrical equipment shall be free from recognized hazards that are likely to cause death or serious physical harm to employees. Safety of equipment shall be determined using the following considerations:

(i) Suitability for installation and use in conformity with the provisions of this subpart. Suitability of equipment for an identified purpose may be evidenced by listing or labeling for that identified purpose.

(ii) Mechanical strength and durability, including, for parts designed to enclose and protect other equipment, the adequacy of the protection thus provided.

(iii) Electrical insulation.

(iv) Heating effects under conditions of use.

(v) Arcing effects.

(vi) Classification by type, size, voltage, current capacity, specific use.

(vii) Other factors which contribute to the practical safeguarding of employees using or likely to come in contact with the equipment.

(2) **Installation and use.** Listed or labeled equipment shall be used or installed in accordance with any instructions included in the listing or labeling.

(c) **Splices.** Conductors shall be spliced or joined with splicing devices suitable for the use or by brazing, welding, or soldering with a fusible metal or alloy. Soldered splices shall first be spliced or joined as to be mechanically and electrically secure without solder and then soldered. All splices and joints and the free ends of conductors shall be covered with an insulation equivalent to that of the conductors or with an insulating device suitable for the purpose.

(d) **Arcing parts.** Parts of electric equipment which in ordinary operation produce arcs, sparks, flames, or molten metal shall be enclosed or separated and isolated from all combustible material.

(e) **Marking.** Electrical equipment may not be used unless the manufacturer's name, trademark, or other descriptive marking by which the organization responsible for the product may be identified is placed on the equipment. Other markings shall be provided giving voltage, current, wattage, or other ratings as necessary. The marking shall be of sufficient durability to withstand the environment involved.

(f) **Identification of disconnecting means and circuits.** Each disconnecting means required by this subpart for motors and appliances shall be legibly marked to indicate its purpose, unless located and arranged so the purpose is evident. Each service, feeder, and branch circuit, at its disconnecting means or overcurrent device, shall be legibly marked to indicate its purpose, unless located and arranged so the purpose is evident. These markings shall be of sufficient durability to withstand the environment involved.

(g) **600 Volts, nominal, or less**

(1) **Working space about electric equipment.** Sufficient access and working space shall be provided and maintained about all electric equipment to permit ready and safe operation and maintenance of such equipment.

(i) **Working clearances.** Except as required or permitted elsewhere in this subpart, the dimension of the working space in the direction of access to live parts operating at 600 volts or less and likely to require examination, adjustment, servicing, or maintenance while alive may not be less than indicated in Table 1. In addition to the dimensions shown in Table 1,

FEDERAL REGULATIONS

Table 6.1 Working Clearances[a]

Nominal voltage to ground	Minimum clear distance for condition[b] (ft)		
	(a)	(b)	(c)
0-150	3	3	3
151-600	3	3½	4

[a]Minimum clear distances may be 2 feet 6 inches for installations built prior to April 16, 1981.
[b]Conditions (a), (b), and (c) are as follows: (a) Exposed live parts on one side and no live or grounded parts on the other side of the working space, or exposed live parts on both sides effectively guarded by suitable wood or other insulating material. Insulated wire or insulated busbars operating at not over 300 volts are not considered live parts. (b) Exposed live parts on one side and grounded parts on the other side. (c) Exposed live parts on both sides of the workspace [not guarded as provided in Condition (a)] with the operator between.

workspace may not be less than 30 inches wide in front of the electric equipment. Distances shall be measured from the live parts if they are exposed, or from the enclosure front or opening if the live parts are enclosed. Concrete, brick, or tile walls are considered to be grounded. Working space is not required in back of assemblies such as dead-front switchboards or motor control centers where there are no renewable or adjustable parts such as fuses or switches on the back and where all connections are accessible from locations other than the back.

(ii) **Clear spaces.** Working space required by this subpart may not be used for storage. When normally enclosed live parts are exposed for inspection or servicing, the working space, if in a passageway or general open space, shall be suitably guarded.

(iii) **Access and entrance to working space.** At least one entrance of sufficient area shall be provided to give access to the working space about electric equipment.

(iv) **Front working space.** Where there are live parts normally exposed on the front of switchboards or motor control centers, the working space in front of such equipment may not be less than 3 feet.

(v) **Illumination.** Illumination shall be provided for all working spaces about service equipment, switchboards, panelboards, and motor control centers installed indoors.

(vi) **Headroom.** The minimum headroom of working spaces about service equipment, switchboards, panelboards, or motor control centers shall be 6 feet 3 inches.

Note: As used in this section, a motor control center is an assembly of one or more enclosed sections having a common power bus and principally containing motor control units.

(2) **Guarding of live parts.**

(i) Except as required or permitted elsewhere in this subpart, live parts of electric equipment operating at 50 volts or more shall be guarded against accidental contact by approved cabinets or other forms of approved enclosures, or by any of the following means:

(A) By location in a room, vault, or similar enclosure that is accessible only to qualified persons.

(B) By suitable permanent, substantial partitions or screens so arranged that only qualified persons will have access to the space within reach of the live parts. Any openings in such partitions or screens shall be so sized and located that persons are not likely to come into accidental contact with the live parts or to bring conducting objects into contact with them.

(C) By location on a suitable balcony, gallery, or platform so elevated and arranged as to exclude unqualified persons.

(D) By elevation of 8 feet or more above the floor or other working surface.

(ii) In locations where electric equipment would be exposed to physical damage, enclosures or guards shall be so arranged and of such strength as to prevent such damage.

(iii) Entrances to rooms and other guarded locations containing exposed live parts shall be marked with conspicuous warning signs forbidding unqualified persons to enter.

(h) **Over 600 volts, nominal.**

(1) **General.** Conductors and equipment used on circuits exceeding 600 volts, nominal, shall comply with all applicable provisions of paragraphs (a) through (g) of this section and with the following provisions which supplement or modify those requirements. The provisions of paragraphs (h) (2), (h) (3), and (h) (4) of this section do not apply to equipment on the supply side of the service conductors.

(2) **Enclosure for electrical installations.** Electrical installations in a vault, room, closet, or in an area surrounded by a wall, screen, or fence, access to which is controlled by lock and

FEDERAL REGULATIONS 135

key or other approved means, are considered to be accessible to qualified persons only. A wall, screen, or fence less than 8 feet in height is not considered to prevent access unless it has other features that provide a degree of isolation equivalent to an 8 foot fence. The entrances to all buildings, rooms, or enclosures containing exposed live parts or exposed conductors operating at over 600 volts, nominal, shall be kept locked or shall be under the observation of a qualified person at all times.

(i) **Installations accessible to qualified persons only.** Electrical installations having exposed live parts shall be accessible to qualified persons only and shall comply with the applicable provisions of paragraph (h) (3) of this section.

(ii) **Installations accessible to unqualified persons.** Electrical installations that are open to unqualified persons shall be made with metal-enclosed equipment or shall be enclosed in a vault or in an area, access to which is controlled by a lock. If metal-enclosed equipment is installed so that the bottom of the enclosure is less than 8 feet above the floor, the door or cover shall be kept locked. Metal-enclosed switchgear, unit substations, transformers, pull boxes, connection boxes, and other similar associated equipment shall be marked with appropriate caution signs. If equipment is exposed to physical damage from vehicular traffic, suitable guards shall be provided to prevent such damage. Ventilating or similar openings in metal-enclosed equipment shall be designed so that foreign objects inserted through these openings will be deflected from energized parts.

(3) **Workspace about equipment.** Sufficient space shall be provided and maintained about electric equipment to permit ready and safe operation and maintenance of such equipment. Where energized parts are exposed, the minimum clear workspace may not be less than 6 feet 6 inches high (measured vertically from the floor or platform), or less than 3 feet wide (measured parallel to the equipment). The depth shall be as required in Table 2. The workspace shall be adequate to permit at least a 90-degree opening of doors or hinged panels.

(i) **Working space.** The minimum clear working space in front of electric equipment such as switchboards, control panels, switches, circuit breakers, motor controllers, relays, and similar equipment may not be less than specified in Table

Table 6.2 Minimum Depth of Clear Working Space in Front of Electric Equipment

Nominal voltage to ground	Conditions[b] (ft)		
	(a)	(b)	(c)
601 to 2,500	3	4	5
2,501 to 9,000	4	5	6
9,001 to 25,000	5	6	9
25,001 to 75kV[a]	6	8	10
Above 75kV[a]	8	10	12

[a]Minimum depth of clear working space in front of electric equipment with a nominal voltage to ground above 25,000 volts may be the same as for 25,000 volts under Conditions (a), (b), and (c) for installations built prior to April 16, 1981.

[b]Conditions (a), (b), and (c) are as follows: (a) Exposed live parts on one side and no live or grounded parts on the other side of the working space, or exposed live parts on both sides effectively guarded by suitable wood or other insulating material. Insulated wire or insulated busbars operating at not over 300 volts are not considered live parts. (b) Exposed live parts on one side and grounded parts on the other side. Concrete, brick, or tile walls will be considered as grounded surfaces. (c) Exposed live parts on both sides of the workspace not guarded as provided in Condition (a) with the operator between.

6.2 unless otherwise specified in this subpart. Distances shall be measured from the live parts if they are exposed, or from the enclosure front or opening if the live parts are enclosed. However, working space is not required in back of equipment such as deadfront switchboards or control assemblies where there are no renewable or adjustable parts (such as fuses or switches) on the back and where all connections are accessible from locations other than the back. Where rear access is required to work on de-energized parts on the back of enclosed equipment, a minimum working space of 30 inches horizontally shall be provided.

(ii) **Illumination.** Adequate illumination shall be provided for all working spaces about electric equipment. The lighting outlets shall be so arranged that persons changing lamps or making repairs on the lighting system will not be endangered by live parts or other equipment. The points of control shall be so located that persons are not likely to come in contact with any live part or moving part of the equipment while turning on the lights.

FEDERAL REGULATIONS

(iii) **Elevation of unguarded live parts.** Unguarded live parts above working space shall be maintained at elevations not less than specified in Table 3.

(4) **Entrance and access to workspace.**

(i) At least one entrance not less than 24 inches wide and 6 feet 6 inches high shall be provided to give access to the working space about electric equipment. On switchboard and control panels exceeding 48 inches in width, there shall be one entrance at each end of such board where practicable. Where bare energized parts at any voltage or insulated energized parts above 600 volts are located adjacent to such entrance, they shall be suitably guarded.

(ii) Permanent ladders or stairways shall be provided to give safe access to the working space around electric equipment installed on platforms, balconies, mezzanine floors, or in attic or roof rooms or spaces.

1910.304 Wiring design and protection.

(a) **Use and identification of grounded and grounding conductors**

(1) **Identification of conductors.** A conductor used as a grounded conductor shall be identifiable and distinguishable from all other conductors. A conductor used as an equipment grounding and distinguishable from all other conductors.

(2) **Polarity of connections.** No grounded conductor may be attached to any terminal or lead so as to reverse designated polarity.

(3) **Use of grounding terminals and devices.** A grounding terminal or grounding-type device on a receptacle, cord connector,

Table 6.3 Elevation of Unguarded Energized Parts Above Working Space

Nominal voltage between phases	Minimum elevation
601 to 7,500	8 feet 6 inches
7,501 to 35,000	9 feet
Over 35 kV	9 feet + 0.37 inches per kV above 35 Kv

[a]Note. Minimum elevation may be 8 feet 0 inches for installations built prior to April 16, 1981, if the nominal voltage between phases is in the range of 601-6600 volts.

or attachment plug may not be used for purposes other than grounding.

(b) **Branch circuits.**

(1) **Ground-fault protection for personnel on construction sites.** The employer shall use either ground-fault circuit interrupters as specified in paragraph (b)(1)(i) of this section or an assured equipment grounding conductor program as specified in paragraph (b)(1)(ii) of this section, to protect employees on construction sites. These requirements are in addition to any other requirements for equipment grounding conductors.

(i) **Ground-fault circuit interrupters.** All 120-volt, single-phase, 15- and 20-ampere receptacle outlets on construction sites, which are not a part of the permanent wiring of the building or structure and which are in use by employees, shall have approved ground-fault circuit interrupters for personnel protection. Receptacles on a two-wire, single-phase portable or vehicle-mounted generator rated not more than 5 kW, where the circuit conductors of the generator are insulated from the generator frame and all other grounded surfaces, need not be protected with ground-fault circuit interrupters.

(ii) **Assured equipment grounding conductor program.** The employer shall establish and implement an assured equipment grounding conductor program on construction sites covering all cord sets, receptacles which are not a part of the permanent wiring of the building or structure, and equipment connected by cord and plug, which are available for use or used by employees. This program shall comply with the following minimum requirements:

(A) A written description of the program, including the specific procedures adopted by the employer, shall be available at the jobsite for inspection and copying by the Assistant Secretary and any affected employee.

(B) The employer shall designate one or more competent persons (as defined in 29 CFR 1926.32(f)) to implement the program.

(C) Each cord set, attachment cap, plug, and receptacle of cord sets, and any equipment connected by cord and plug, except cord sets and receptacles which are fixed and not exposed to damage, shall be visually inspected before each day's use for external defects, such as deformed or missing pins or insulation damage, and for indication of possible internal damage. Equipment found damaged or defective may not be used until repaired.

(D) The following tests shall be performed on all cord sets, receptacles which are not a part of the permanent wiring of the building or structure, and cord- and plug-connected equipment required to be grounded:

(1) All equipment grounding conductors shall be tested for continuity and shall be electrically continuous.

(2) Each receptacle and attachment cap or plug shall be tested for correct attachment of the equipment grounding conductor. The equipment grounding conductor shall be connected to its proper terminal.

(E) All required tests shall be performed as follows:

(1) Before first use;

(2) Before equipment is returned to service following any repairs;

(3) Before equipment is used after any incident which can be reasonably suspected to have caused damage (for example, when a cord set is run over); and

(4) At intervals not to exceed 3 months, except that cord sets and receptacles which are fixed and not exposed to damage shall be tested at intervals not exceeding 6 months.

(F) The employer may not make available or permit the use by employees of any equipment which has not met the requirements of this paragraph (b)(1)(ii) of this section.

(G) Tests performed as required in this paragraph shall be recorded. This test record shall identify each receptacle, cord set, and cord- and plug-connected equipment that passed the test, and shall indicate the last date it was tested or the interval for which it was tested. This record shall be kept by means of logs, color coding, or other effective means, and shall be maintained until replaced by a more current record. The record shall be made available on the jobsite for inspection by the Assistant Secretary and any affected employee.

(2) **Outlet devices.** Outlet devices shall have an ampere rating not less than the load to be served.

(c) **Outside conductors, 600 volts, nominal, or less.** Paragraphs (c)(1), (c)(2), (c)(3), and (c)(4) of this section apply to branch circuit, feeder, and service conductors rated 600 volts, nominal, or less and run outdoors as open conductors. Paragraph (c)(5) applies to lamps installed under such conductors.

(1) **Conductors on poles.** Conductors supported on poles shall provide a horizontal climbing space not less than the following:

(i) Power conductors below communication conductors—30 inches.

(ii) Power conductors alone or above communication conductors: 300 volts or less—24 inches; more than 300 volts—30 inches.

(iii) Communication conductors below power conductors with power conductors 300 volts or less—24 inches; more than 300 volts—30 inches.

(2) **Clearance from ground.** Open conductors shall conform to the following minimum clearances:

(i) 10 feet—above finished grade, sidewalks, or from any platform or projection from which they might be reached.

(ii) 12 feet—over areas subject to vehicular traffic other than truck traffic.

(iii) 15 feet—over areas other than those specified in paragraph (c)(2)(iv) of this section that are subject to truck traffic.

(iv) 18 feet—over public streets, alleys, roads, and driveways.

(3) **Clearance from building openings.** Conductors shall have a clearance of at least 3 feet from windows, doors, porches, fire escapes, or similar locations. Conductors run above the top level of a window are considered to be out of reach from that window and, therefore, do not have to be 3 feet away.

(4) **Clearance over roofs.** Conductors shall have a clearance of not less than 8 feet from the highest point of roofs over which they pass, except for the following:

(k) Where the voltage between conductors is 300 volts or less and the roof has a slope of not less than 4 inches in 12, the clearance from roofs shall be at least 3 feet, or

(ii) Where the voltage between conductors is 300 volts or less and the conductors do not pass over more than 4 feet of the overhang portion of the roof and they are terminated at a through-the-roof raceway or approved support, the clearance from roofs shall be at least 18 inches.

(5) **Location of outdoor lamps.** Lamps for outdoor lighting shall be located below all live conductors, transformers, or other electric equipment, unless such equipment is controlled by a disconnecting means that can be locked in the open position or unless adequate clearances or other safeguards are provided for relamping operations.

(d) **Services.**

(1) **Disconnecting means.**

(i) **General.** Means shall be provided to disconnect all conductors in a building or other structure from the service-entrance conductors. The disconnecting means shall plainly indicate whether it is in the open or closed position and shall be installed at a readily accessible location nearest the point of entrance of the service-entrance conductors.

(ii) **Simultaneous opening of poles.** Each service disconnecting means shall simultaneously disconnect all ungrounded conductors.

(2) **Services over 600 volts, nominal.** The following additional requirements apply to services over 600 volts, nominal.

(i) **Guarding.** Service-entrance conductors installed as open wires shall be guarded to make them accessible only to qualified persons.

(ii) **Warning signs.** Signs warning of high voltage shall be posted where other than qualified employees might come in contact with live parts.

(e) **Overcurrent protection.**

(1) **600 volts, nominal, or less.** The following requirements apply to overcurrent protection of circuits rated 600 volts, nominal, or less.

(i) **Protection of conductors and equipment.** Conductors and equipment shall be protected from overcurrent in accordance with their ability to conduct current safely.

(ii) **Grounded conductors.** Excepts for motor running overload protection, overcurrent devices may not interrupt the continuity of the grounded conductor unless all conductors of the circuit are opened simultaneously.

(iii) **Disconnecting of fuses and thermal cutouts.** Except for service fuses, all cartridge fuses which are accessible to other than qualified persons and all fuses and thermal cutouts on circuits over 150 volts to ground shall be provided with disconnecting means. This disconnecting means shall be installed so that the fuse or thermal cutout can be disconnected from its supply without disrupting service to equipment and circuits unrelated to those protected by the overcurrent device.

(iv) **Location in or on premises.** Overcurrent devices shall be readily accessible to each employee or authorized building management personnel. These overcurrent devices may not be located where they will be exposed to physical damage nor in the vicinity of easily ignitible material.

(v) **Arcing or suddenly moving parts.** Fuses and circuit breakers shall be so located or shielded that employees will not be burned or otherwise injured by their operation.

(vi) **Circuit breakers.**

(A) Circuit breakers shall clearly indicate whether they are in the open (off) or closed (on) position.

(B) Where circuit breaker handles on switchboards are operated vertically rather than horizontally or rotationally, the up position of the handle shall be the closed (on) position. (See 1910.302(b)(3).)

(C) If used as switches in 120-volt, fluorescent lighting circuits, circuit breakers shall be approved for the purpose and marked "SWD." (See 1910.302(b)(3).)

(2) **Over 600 volts, nominal.** Feeders and branch circuits over 600 volts, nominal, shall have short-circuit protection.

(f) **Grounding.** Paragraphs (f)(1) through (f)(7) of this section contain grounding requirements for systems, circuits, and equipment.

(1) **Systems to be grounded.** The following systems which supply premises wiring shall be grounded:

(i) All 3-wire DC systems shall have their neutral conductor grounded.

(ii) Two-wire DC systems operating at over 50 volts through 300 volts between conductors shall be grounded unless:

(A) They supply only industrial equipment in limited areas and are equipped with a ground detector; or

(B) They are rectifier-derived from an AC system complying with paragraph (f)(1)(iii), (f)(1)(iv), and (f)(1)(v) of this section; or

(C) They are fire-protective signaling circuits having a maximum current of 0.030 amperes.

(iii) AC circuits of less than 50 volts shall be grounded if they are installed as overhead conductors outside of buildings or if they are supplied by transformers and the transformer primary supply system is ungrounded or exceeds 150 volts to ground.

(iv) AC systems of 50 volts to 1000 volts shall be grounded under any of the following conditions, unless exempted by paragraph (f)(1)(v) of this section:

(A) If the system can be so grounded that the maximum voltage to ground on the ungrounded conductors does not exceed 150 volts;

(B) If the system is nominally rated 480Y/277 volt, 3-phase, 4-wire in which the neutral is used as a circuit conductor;

(C) If the system is nominally rated 240/120 volt, 3-phase, 4-wire in which the midpoint of one phase is used as a circuit conductor; or

(D) If a service conductor is uninsulated.

(v) AC systems of 50 volts to 1000 volts are not required to be grounded under any of the following conditions:

(A) If the system is used exclusively to supply industrial electric furnaces for melting, refining, tempering, and the like.

(B) If the system is separately derived and is used exclusively for rectifiers supplying only adjustable speed industrial drives.

(C) If the system is separately derived and is supplied by a transformer that has a primary voltage rating less than 1000 volts, provided all of the following conditions are met:

(1) The system is used exclusively for control circuits,

(2) The conditions of maintenance and supervision assure that only qualified persons will service the installation,

(3) Continuity of control power is required, and

(4) Ground detectors are installed on the control system.

(D) If the system is an isolated power system that supplies circuits in health care facilities.

(2) **Conductors to be grounded.** For AC premises wiring systems, the identified conductor shall be grounded.

(3) **Grounding connections.**

(i) For a grounded system, a grounding electrode conductor shall be used to connect both the equipment grounding conductor and the grounded circuit conductor to the grounding electrode. Both the equipment grounding conductor and the grounding electrode conductor shall be connected to the grounded circuit conductor on the supply side of the service disconnecting means, or on the supply side of the system disconnecting means or overcurrent devices if the system is separately derived.

(ii) For an ungrounded service-supplied system, the equipment grounding conductor shall be connected to the grounding electrode conductor at the service equipment. For an ungrounded separately derived system, the equipment grounding conductor shall be connected to the grounding electrode conductor at, or ahead of, the system disconnecting means or overcurrent devices.

(iii) On extensions of existing branch circuits which do not have an equipment grounding conductor, grounding-type receptacles may be grounded to a grounded cold water pipe near the equipment.

(4) **Grounding path.** The path to ground from circuits, equipment, and enclosures shall be permanent and continuous.

(5) **Supports, enclosures, and equipment to be grounded.**

(k) **Supports and enclosures for conductors.** Metal cable trays, metal raceways, and metal enclosures for conductors shall be grounded, except that:

(A) Metal enclosures such as sleeves that are used to protect cable assemblies from physical damage need not be grounded; or

(B) Metal enclosures for conductors added to existing installations of open wire, knob-and-tube wiring, and nonmetallic-sheathed cable need not be grounded if all of the following conditions are met:

(1) Runs are less than 25 feet;

(2) Enclosures are free from probable contact with ground, grounded metal, metal laths, or other conductive materials; and

(3) Enclosures are guarded against employee contact.

(ii) **Service equipment enclosures.** Metal enclosures for service equipment shall be grounded.

(iii) **Frames of ranges and clothes dryers.** Frames of electric ranges, wall-mounted ovens, counter-mounted cooking units, clothes dryers, and metal outlet or junction boxes which are part of the circuit for these appliances shall be grounded.

(iv) **Fixed equipment.** Exposed non-current-carrying metal parts of fixed equipment which may become energized shall be grounded under any of the following conditions:

(A) If within 8 feet vertically or 5 feet horizontally of ground or grounded metal objects and subject to employee contact.

(B) If located in a wet or damp location and not isolated.

(C) If in electrical contact with metal.

(D) If in a hazardous (classified) location.

(E) If supplied by a metal-clad, metal-sheathed, or grounded metal raceway wiring method.

(F) If equipment operates with any terminal at over 150 volts to ground; however, the following need not be grounded:

(1) Enclosures for switches or circuit breakers used for other than service equipment and accessible to qualified persons only;

(2) Metal frames of electrically heated appliances which are permanently and effectively insulated from ground; and

(3) The cases of distribution apparatus such as transformers and capacitors mounted on wooden poles at a height exceeding 8 feet above ground or grade level.

(v) **Equipment connected by cord and plug.** Under any of the conditions described in paragraphs (f)(5)(v)(C) of this section, exposed non-current-carrying metal parts of cord- and plug-connected equipment which may become energized shall be grounded.

(A) If in hazardous (classified) locations (see 1910.307).

(B) If operated at over 150 volts to ground, except for guarded motors and metal frames of electrically heated appliances if the appliance frames are permanently and effectively insulated from ground.

(C) If the equipment is of the following types:

(1) Refrigerators, freezers, and air conditioners;

(2) Clothes-washing, clothes-drying, and dishwashing machines, sump pumps, and electrical aquarium equipment;

(3) Hand-held motor-operated tools;

(4) Motor-operated appliances of the following types: hedge clippers, lawn mowers, snow blowers, and wet scrubbers;

(5) Cord- and plug-connected appliances used in damp or wet locations or by employees standing on the ground or on metal floors or working inside of metal tanks or boilers;

(6) Portable and mobile X-ray and associated equipment;

(7) Tools likely to be used in wet and conductive locations; and

(8) Portable hand lamps.

Tools likely to be used in wet and conductive locations need not be grounded if supplied through an isolating transformer with an ungrounded secondary of not over 50 volts. Listed or labeled portable tools and appliances protected by an approved system of double insulation, or its equivalent, need not be grounded. If such a system is employed, the equipment shall be distinctively marked to indicate that the tool or appliance utilizes an approved system of double insulation.

(vi) **Nonelectrical equipment.** The metal parts of the following nonelectrical equipment shall be grounded: frames and tracks of electrically operated cranes; frames of nonelectrically driven elevator cars to which electric conductors are attached; hand operated metal shifting ropes or cables or electric elevators, and metal partitions, grill work, and similar metal enclosures around equipment of over 750 volts between conductors.

(6) **Methods of grounding fixed equipment.**

(i) Non-current-carrying metal parts of fixed equipment, if required to be grounded by this subpart, shall be grounded by an equipment grounding conductor which is contained within the same raceway, cable, or cord, or runs with or encloses the circuit conductors. For DC circuits only, the equipment grounding conductor may be run separately from the circuit conductors.

(ii) Electric equipment is considered to be effectively grounded if it is secured to, and in electrical contact with, a metal rack or structure that is provided for its support and the metal rack or structure is grounded by the method specified for the non-current-carrying metal parts of fixed equipment in paragraph (f)(6)(i) of this section. For installations made before April 16, 1981, only, electric equipment is also considered to be effectively grounded if it is secured to, and in metallic contact with, the grounded structural metal frame of a building. Metal car frames supported by metal hoisting cables attached to or running over metal sheaves or drums of grounded elevator machines are also considered to be effectively grounded.

(7) **Grounding of systems and circuits of 1000 volts and over (high voltage).**

(i) General. If high voltage systems are grounded, they shall comply with all applicable provisions of paragraphs (f)(1) through (f)(6) of this section as supplemented and modified by this paragraph (f)(7).

(ii) **Grounding of systems supplying portable or mobile equipment.** (See 1910.302(b)(3).) Systems supplying portable or mobile high voltage equipment, other than substations installed on a temporary basis, shall comply with the following:

(A) Portable and mobile high voltage equipment shall be supplied from a system having its neutral grounded through an impedance. If a delta-connected high voltage system is used to supply the equipment, a system neutral shall be derived.

(B) Exposed non-current-carrying metal parts of portable and mobile equipment shall be connected by an equipment grounding conductor to the point at which the system neutral impedance is grounded.

(C) Ground-fault detection and relaying shall be provided to de-energize automatically any high voltage system component which has developed a ground fault. The continuity of the equipment grounding conductor shall be continuously monitored so as to de-energize automatically the high voltage feeder to the portable equipment upon loss of continuity of the equipment grounding conductor.

(D) The grounding electrode to which the portable or mobile equipment system neutral impedance is connected shall be isolated from and separated in the ground by at least 20 feet from any other system or equipment grounding electrode, and there shall be no direct connection between the grounding electrodes, such as buried pipe, fence, etc.

(iii) **Grounding of equipment.** All non-current-carrying metal parts of portable equipment and fixed equipment including their associated fences, housings, enclosures, and supporting structures shall be grounded. However, equipment which is guarded by location and isolated from ground need not be grounded. Additionally, pole-mounted distribution apparatus at a height exceeding 8 feet above ground or grade level need not be grounded.

1910.305 Wiring methods, components, and equipment for general use.

(a) **Wiring methods.** The provisions of this section do not apply to the conductors that are an integral part of factory-assembled equipment.

(1) **General requirements.**

(i) **Electrical continuity of metal raceways and enclosures.** Metal raceways, cable armor, and other metal enclosures for conductors shall be metallically joined together into a continuous electric conductor and shall be so connected to all boxes, fittings, and cabinets as to provide effective electrical continuity.

(ii) **Wiring in ducts.** No wiring systems of any type shall be installed in ducts used to transport dust, loose stock, or flammable vapors. No wiring system of any type may be installed in any duct used for vapor removal or for ventilation of commercial-type cooking equipment, or in any shaft containing only such ducts.

(2) **Temporary wiring.** Temporary electrical power and

lighting wiring methods may be of a class less than would be required for a permanent installation. Except as specifically modified in this paragraph, all other requirements of this subpart for permanent wiring shall apply to temporary wiring installations.

(i) **Uses permitted, 600 volts, nominal, or less.** Temporary electrical power and lighting installations 600 volts, nominal, or less may be used only as follows:

(A) During and for remodeling, maintenance, repair, or demolition of buildings, structures, or equipment, and similar activities;

(B) For experimental or development work, and

(C) For a period not to exceed 90 days for Christmas decorative lighting, carnivals, and similar purposes.

(ii) **Uses permitted, over 600 volts, nominal.** Temporary wiring over 600 volts, nominal, may be used only during periods of tests, experiments, or emergencies.

(iii) **General requirements for temporary wiring.**

(A) Feeders shall originate in an approved distribution center. The conductors shall be run as multiconductor cord or cable assemblies, or, where not subject to physical damage, they may be run as open conductors on insulators not more than 10 feet apart.

(B) Branch circuits shall originate in an approved power outlet or panelboard. Conductors shall be multiconductor cord or cable assemblies or open conductors. If run as open conductors, they shall be fastened at ceiling height every 10 feet. No branch-circuit conductor may be laid on the floor. Each branch circuit that supplies receptacles or fixed equipment shall contain a separate equipment grounding conductor if run as open conductors.

(C) Receptacles shall be of the grounding type. Unless installed in a complete metallic raceway, each branch circuit shall contain a separate equipment grounding conductor and all receptacles shall be electrically connected to the grounding conductor.

(D) No bare conductors nor earth returns may be used for the wiring of any temporary circuit.

(E) Suitable disconnecting switches or plug connectors shall be installed to permit the disconnection of all ungrounded conductors of each temporary circuit.

(F) Lamps for general illumination shall be protected from accidental contact or breakage. Protection shall be provided by elevation of at least 7 feet from normal working surface or by a suitable fixture or lampholder with a guard.

(G) Flexible cords and cables shall be protected from accidental damage. Sharp corners and projections shall be avoided. Where passing through doorways or other pinch points, flexible cords and cables shall be provided with protection to avoid damage.

(3) **Cable trays.**

(i) **Uses permitted.**

Only the following may be installed in cable tray systems:

(1) Mineral-insulated metal-sheathed cable (Type MI);

(2) Armored cable (Type AC);

(3) Metal-clad cable (Type MC);

(4) Power-limited tray cable (Type PLTC);

(5) Nonmetallic-sheathed cable (Type NM or NMC);

(6) Shielded nonmetallic-sheathed cable (Type SNM);

(7) Multiconductor service-entrance cable (Type SE or USE);

(8) Multiconductor underground feeder and branch-circuit cable (Type UF);

(9) Power and control tray cable (Type TC);

(10) Other factory-assembled, multiconductor control, signal, or power cables which are specifically approved for installation in cable trays; or

(11) Any approved conduit or raceway with its contained conductors.

In industrial establishments only, where conditions of maintenance and supervision assure that only qualified persons will service the installed cable tray system, the following cables may also be installed in ladder, ventilated trough, or 4 inch ventilated channel-type cable trays:

(1) Single conductor cables which are 250 MCM or larger and are Types RHH, RHW, MV, USE, or THW, and other 250 MCM or larger single conductor cables if specifically approved for installation in cable trays. Where exposed to direct rays of the sun, cables shall be sunlight-resistant.

(2) Type MV cables, where exposed to direct rays of the sun, shall be sunlight-resistant.

Cable trays in hazardous (classified) locations shall contain only the cable types permitted in such locations.

(ii) **Uses not permitted.** Cable tray systems may not be used in hoistways or where subjected to severe physical damage.

(f) **Open wiring on insulators.**

(i) **Uses permitted.** Open wiring on insulators is only permitted on systems of 600 volts, nominal, or less for industrial or agricultural establishments and for services.

(ii) **Conductor supports.** Conductors shall be rigidly supported on noncombustible, nonabsorbent insulating materials and may not contact any other objects.

(iii) **Flexible nonmetallic tubing.** In dry locations where not exposed to severe physical damage, conductors may be separately enclosed in flexible nonmetallic tubing. The tubing shall be in continuous lengths not exceeding 15 feet and secured to the surface by straps at intervals not exceeding 4 feet 6 inches.

(iv) **Through walls, floors, wood cross members, etc.** Open conductors shall be separated from contact with walls, floors, wood cross members, or partitions through which they pass by tubes or bushings of noncombustible, nonabsorbent insulating material. If the bushing is shorter than the hole, a waterproof sleeve of nonconductive material shall be inserted in the hole and an insulating bushing slipped into the sleeve at each end in such a manner as to keep the conductors absolutely out of contact with the sleeve. Each conductor shall be carried through a separate tube or sleeve.

(v) **Protection from physical damage.** Conductors within 7 feet from the floor are considered exposed to physical damage. Where open conductors cross ceiling joints and wall studs and are exposed to physical damage, they shall be protected.

(b) **Cabinets, boxes, and fittings.**

(1) **Conductors entering boxes, cabinets, or fittings.** Conductors entering boxes, cabinets, or fittings shall also be protected from abrasion, and openings through which conductors enter shall be effectively closed. Unused openings in cabinets, boxes, and fittings shall be effectively closed.

(2) **Covers and canopies.** All pull boxes, junction boxes, and fittings shall be provided with covers approved for the purpose. If metal covers are used, they shall be grounded. In completed installations, each outlet box shall have a cover, faceplate, or fixture canopy. Covers of outlet boxes having holes through which flexible cord pendants pass shall be provided with bushings designed for the purpose or shall have smooth, well-rounded surfaces on which the cords may bear.

(3) **Pull and junction boxes for systems over 600 volts, nominal.** In addition to other requirements in this section for pull and junction boxes, the following shall apply to these boxes for systems over 600 volts, nominal:

(i) Boxes shall provide a complete enclosure for the contained conductors or cables.

(ii) Boxes shall be closed by suitable covers securely fastened in place. Underground box covers that weigh over 100 pounds meet this requirement. Covers for boxes shall be permanently marked "HIGH VOLTAGE." The marking shall be on the outside of the box cover and shall be readily visible and legible.

(c) **Switches.**

(1) **Knife switches.** Single-throw knife switches shall be so connected that the blades are dead when the switch is in the open position. Single-throw knife switches shall be so placed that gravity will not tend to close them. Single-throw knife switches approved for use in the inverted position shall be provided with a locking device that will ensure that the blades remain in the open position when so set. Double-throw knife switches may be mounted so that the throw will be either vertical or horizontal. However, if the throw is vertical, a locking device shall be provided to ensure that the blades remain in the open position when so set.

(2) **Faceplates for flush-mounted snap switches.** Flush snap switches that are mounted in ungrounded metal boxes and

located within reach of conducting floors or other conducting surfaces shall be provided with faceplates of nonconducting, noncombustible material.

(d) **Switchboards and panelboards.** Switchboards that have any exposed live parts shall be located in permanently dry locations and accessible only to qualified persons. Panelboards shall be mounted in cabinets, cutout boxes, or enclosures approved for the purpose and shall be dead front. However, panelboards other than the dead front externally-operable type are permitted where accessible only to qualified persons. Exposed blades of knife switches shall be dead when open.

(e) **Enclosures for damp or wet locations.**

(1) Cabinets, cutout boxes, fittings, boxes, and panelboard enclosures in damp or wet locations shall be installed so as to prevent moisture or water from entering and accumulating within the enclosures. In wet locations, the enclosures shall be weatherproof.

(2) Switches, circuit breakers, and switchboards installed in wet locations shall be enclosed in weatherproof enclosures.

(f) **Conductors for general wiring.**

All conductors used for general wiring shall be insulated unless otherwise permitted in this Subpart. The conductor insulation shall be of a type that is approved for the voltage, operating temperature, and location of use. Insulated conductors shall be distinguishable by appropriate color or other suitable means as being grounded conductors, ungrounded conductors, or equipment grounding conductors.

(g) **Flexible cords and cables.**

(1) **Use of flexible cords and cables.**

(i) Flexible cords and cables shall be approved and suitable for conditions of use and location. Flexible cords and cables shall be used only for the following:

(A) Pendants;

(B) Wiring of fixtures;

(C) Connection of portable lamps or appliances;

(D) Elevator cables;

(E) Wiring of cranes and hoists;

(F) Connection of stationary equipment to facilitate their frequent interchange;

(G) Prevention of the transmission of noise or vibration;

(H) Appliances where the fastening means and mechanical connections are designed to permit removal for maintenance and repair; or

(i) Data processing cables approved as a part of the data processing system.

(ii) If used as permitted in paragraphs (g)(1)(i)(c), (g)(1)(i)(f), or (g)(1)(i)(h) of this section, the flexible cord shall be equipped with an attachment plug and shall be energized from an approved receptacle outlet.

(iii) Unless specifically permitted in paragraph (g)(1)(i) of this section, flexible cords and cables may not be used:

(A) As a substitute for the fixed wiring of a structure;

(B) Where run through holes in walls, ceilings, or floors;

(C) Where run through doorways, windows, or similar openings;

(D) Where attached to building surfaces; or

(E) Where concealed behind building walls, ceilings, or floors.

(iv) Flexible cords used in show windows and showcases shall be Type S, SO, SJ, SJO, STO, ST, SJT, SJTO, of AFS except for the wiring of chain-supported lighting fixtures and supply cords for portable lamps and other merchandise being displayed or exhibited.

(2) **Identification, splices, and terminations.**

(i) A conductor of a flexible cord or cable that is used as a grounded conductor or an equipment grounding conductor shall be distinguishable from other conductors. Types SJ, SJO, SJT, SJTO, S, SO, ST, and STO shall be durably marked on the surface with the type designation, size, and number of conductors.

(ii) Flexible cords shall be used only in continuous lengths without splice or tap. Hard service flexible cords No. 12 or larger may be repaired if spliced so that the splice retains the insulation, outer sheath properties, and usage characteristics of the cord being spliced.

(iii) Flexible cords shall be connected to devices and fittings so that strain relief is provided which will prevent pull from being directly transmitted to joints or terminal screws.

(h) **Portable cables over 600 volts, nominal.** Multiconductor portable cable for use in supplying power to portable or mobile equipment at over 600 volts, nominal, shall consist of No. 8 or larger conductors employing flexible stranding. Cables operated at over 2,000 volts shall be shielded for the purpose of confining the voltage stresses to the insulation. Grounding conductors shall be provided. Connectors for these cables shall be of a locking type with provisions to prevent their opening or closing while energized. Strain relief shall be provided at connections and terminations. Portable cables may not be operated with splices unless the splices are of the permanent molded, vulcanized, or other approved type. Termination enclosures shall be suitably marked with a high voltage hazard warning, and terminations shall be accessible only to authorized and qualified personnel.

(i) **Fixture wires.**

(1) **General.** Fixture wires shall be approved for the voltage, temperature, and location of use. A fixture wire which is used as a grounded conductor shall be identified.

(2) **Uses permitted.** Fixture wires may be used as follows:

(i) For installation in lighting fixtures and in similar equipment where enclosed or protected and not subject to bending or twisting in use; or

(ii) For connecting lighting fixtures to the branch-circuit conductors supplying the fixtures.

(3) **Uses not permitted.** Fixture wires may not be used as branch-circuit conductors except as permitted for Class 1 power limited circuits.

(j) **Equipment for general use.**

(1) **Lighting fixtures, lampholders, lamps, and receptacles.**

(i) Fixtures, lampholders, lamps, rosettes, and receptacles may have no live parts normally exposed to employee contact. However, rosettes and cleat-type lampholders and receptacles located at least 8 feet above the floor may have exposed parts.

(ii) Handlamps of the portable type supplied through flexible cords shall be equipped with a handle of molded composition or other material approved for the purpose, and a substantial guard shall be attached to the lampholder or the handle.

(iii) Lampholders of the screw-shell type shall be installed for use as lampholders only. Lampholders installed in wet or damp locations shall be of the weatherproof type.

(iv) Fixtures installed in wet or damp locations shall be approved for the purpose and shall be so constructed or installed that water cannot enter or accumulate in wireways, lampholders, or other electrical parts.

(2) **Receptacles, cord connectors, and attachment plugs (caps).**

(i) Receptacles, cord connectors, and attachment plugs shall be constructed so that no receptacle or cord connector will accept an attachment plug with a different voltage or current rating than that for which the device is intended. However, a 20-ampere T-slot receptacle or cord connector may accept a 15-ampere attachment plug of the same voltage rating.

(ii) A receptacle installed in a wet or damp location shall be suitable for the location.

(3) **Appliances.**

(i) Appliances, other than those in which the current-carrying parts at high temperatures are necessarily exposed, may have no live parts normally exposed to employee contact.

(ii) A means shall be provided to disconnect each appliance.

(iii) Each appliance shall be marked with its rating in volts and amperes or volts and watts.

(4) **Motors.** This paragraph applies to motors, motor circuits, and controllers.

(i) **In sight form.** If specified that one piece of equipment shall be "in sight from" another piece of equipment, one shall be visible and not more than 50 feet from the other.

(ii) **Disconnecting means.**

(A) A disconnecting means shall be located in sight from the controller location. However, a single disconnecting means may be located adjacent to a group of coordinated controllers mounted adjacent to each other on a multi-motor continuous process machine. The controller disconnecting means for motor branch circuits over 600 volts, nominal, may be out of sight of the controller, if the controller is marked with a warning label giving the location and identification of the disconnecting means which is to be locked in the open position.

(B) The disconnecting means shall disconnect the motor and the controller from all ungrounded supply conductors and shall be so designed that no pole can be operated independently.

(C) If a motor and the driven machinery are not in sight from the controller location, the installation shall comply with one of the following conditions:

(1) The controller disconnecting means shall be capable of being locked in the open position.

(2) A manually operable switch that will disconnect the motor from its source of supply shall be placed in sight from the motor location.

(D) The disconnecting means shall plainly indicate whether it is in the open (off) or closed (on) position.

(E) The disconnecting means shall be readily accessible. If more than one disconnect is provided for the same equipment, only one need be readily accessible.

(F) An individual disconnecting means shall be provided for each motor, but a single disconnecting means may be used for a group of motors under any one of the following conditions:

(1) If a number of motors drive special parts of a single machine or piece of apparatus, such as a metal or woodworking machine, crane, or hoist;

(2) If a group of motors is under the protection of one set of branch-circuit protective devices; or

(3) If a group of motors is in a single room in sight from the location of the disconnecting means.

(iii) **Motor overload, short-circuit, and ground-fault protection.** Motors, motor-control apparatus, and motor branch-circuit conductors shall be protected against over heating due to motor overloads or failure to start, and against short circuits or ground faults. These provisions shall not require overload protection that will stop a motor where a shutdown is likely to introduce additional or increased hazards, as in the case of fire pumps, or where continued operation of a motor is necessary for a safe shutdown of equipment or process and motor overload sensing devices are connected to a supervised alarm.

(iv) **Protection of live parts—all voltages.**

(A) Stationary motors having commutators, collectors, and brush rigging located inside of motor end brackets and not conductively connected to supply circuits operating at more than 150 volts to ground need not have such parts guarded. Exposed live parts of motors and controllers operating at 50 volts or more between terminals shall be guarded against accidental contact by any of the following:

(1) By Installation in a room or enclosure that is accessible only to qualified persons;

(2) By installation on a suitable balcony, gallery, or platform, so elevated and arranged as to exclude unqualified persons; or

(3) By elevation 8 feet or more above the floor.

(B) Where live parts of motors or controllers operating at over 150 volts to ground are guarded against accidental contact only by location, and where adjustment or other attendance may be necessary during the operation of the apparatus, suitable insulating mats or platforms shall be provided so that the attendant cannot readily touch live parts unless standing on the mats or platforms.

(5) **Transformers.**

(i) The following paragraphs cover the installation of all transformers except the following:

(A) Current transformers;

(B) Dry-type transformers installed as a component part of other apparatus;

(D) Transformers used with Class 2 and Class 3 circuits, sign and outline lighting, electric discharge lighting, and power-limited fire-protective signalling circuits; and

(E) Liquid-filled or dry-type transformers used for research, development, or testing, where effective safeguard arrangements are provided.

(ii) The operating voltage of exposed live parts of transformer installations shall be indicated by warning signs or visible markings on the equipment or structure.

(iii) Dry-type, high fire point liquid-insulated, and askarel-insulated transformers installed indoors and rated over 35kV shall be in a vault.

(iv) If they present a fire hazard to employees, oilinsulated transformers installed indoors shall be in a vault.

(v) Combustible material, combustible buildings and parts of buildings, fire escapes, and door and window openings shall be safeguarded from fires which may originate in oilinsulated transformers attached to or adjacent to a building or combustible material.

(vi) Transformer vaults shall be constructed so as to contain fire and combustible liquids within the vault and to prevent unauthorized access. Locks and latches shall be so arranged that a vault door can be readily opened from the inside.

(vii) Any pipe or duct system foreign to the vault installation may not enter or pass through a transformer vault.

(viii) Materials may not be stored in transformer vaults.

(6) **Capacitors.**

(i) All capacitors, except surge capacitors or capacitors included as a component part of other apparatus, shall be provided with an automatic means of draining the stored charge after the capacitor is disconnected from its source of supply.

(ii) Capacitors rated over 600 volts, nominal, shall comply with the following additional requirements:

(A) Isolating or disconnecting switches (with no interrupting rating) shall be interlocked with the load interrupting device or shall be provided with prominently displayed caution signs to prevent switching load current.

(B) For series capacitors (see 1910.302(b)(3)), the proper switching shall be assured by use of at least one of the following:

(1) Mechanically sequenced isolating and bypass switches,

(2) Interlocks, or

(3) Switching procedure prominently displayed at the switching location.

(7) **Storage batteries.** Provisions shall be made for sufficient diffusion and ventilation of gasses from storage batteries to prevent the accumulation of explosive mixtures.

1910.306 Specific purpose equipment and installations

(a) **Electric signs and outline lighting**

(1) **Disconnecting means.** Signs operated by electronic or electromechanical controllers located outside the sign shall have a disconnecting means located inside the controller enclosure or within sight of the controller location, and it shall be capable of being locked in the open position. Such disconnecting means shall have no pole that can be operated independently, and it shall open all ungrounded conductors that supply the controller and sign. All other signs, except the portable type, and all outline lighting installations shall have an externally operable disconnecting means which can open all ungrounded conductors and is within the sight of the sign or outline lighting it controls.

(2) Doors or covers giving access to uninsulated parts of indoor signs or outline lighting exceeding 600 volts and accessible to other than qualified persons shall either be provided with interlock switches to disconnect the primary circuit or shall be so fastened that the use of other than ordinary tools will be necessary to open them.

(b) **Cranes and hoists.** This paragraph applies to the installation of electric equipment and wiring used in connection with cranes, monorail hoists, hoists, and all runways.

(1) **Disconnecting means.** A readily accessible disconnecting means

(i) Shall be provided between the runway contact conductors and the power supply.

(ii) Another disconnecting means, capable of being locked in the open position, shall be provided in the leads from the runway contact conductors or other power supply on any crane or monorail hoist.

(A) If this additional disconnecting means is not readily accessible from the crane or monorail hoist operating station, means shall be provided at the operating station to open the power circuit to all motors of the crane or monorail hoist.

(B) The additional disconnect may be omitted if a monorail hoist or hand-propelled crane bridge installation meets all of the following:

(1) The unit is floor controlled;

(2) The unit is within view of the power supply disconnecting means; and

(3) No fixed work platform has been provided for servicing the unit.

(2) **Control.** A limit switch or other device shall be provided to prevent the load block from passing the safe upper limit of travel of any hoisting mechanism.

(3) **Clearance.** The dimension of the working space in the direction of access to live parts which may require examination, adjustment, servicing, or maintenance while alive shall be a minimum of 2 feet 6 inches. Where controls are enclosed in cabinets, the door(s) shall either open at least 90 degrees or be removable.

(c) **Elevators, dumbwaiters, escalators, and moving walks.**

(1) **Disconnecting means.** Elevators, dumbwaiters, escalators, and moving walks shall have a single means for disconnecting all ungrounded main power supply conductors for each unit.

(2) **Warning signs.** If interconnections between control panels are necessary for operation of the system on a multicar installation that remains energized from a source other than the disconnecting means, a warning sign shall be mounted on or adjacent to the disconnecting means. The sign shall be clearly legible and shall read "Warning—Parts of the control panel are not de-energized by this switch." (See 1910.302(b)(3).)

(3) **Control panels.** If control panels are not located in the same space as the drive machine, they shall be located in

cabinets with doors or panels capable of being locked closed.

(d) **Electric welders—disconnecting means.**

(1) A disconnecting means shall be provided in the supply circuit for each motor-generator arc welder, and for each AC transformer and DC rectifier arc welder which is not equipped with a disconnect mounted as an intergral part of the welder.

(2) A switch or circuit breaker shall be provided by which each resistance welder and its control equipment can be isolated from the supply circuit. The ampere rating of this disconnecting means may not be less than the supply conductor ampacity.

(e) **Data processing systems—disconnecting means.** A disconnecting means shall be provided to disconnect the power to all electronic equipment in data processing or computer rooms. This disconnecting means shall be controlled from locations readily accessible to the operator at the principal exit doors. There shall also be a similar disconnecting means to disconnect the air conditioning system serving this area.

(f) **X-Ray equipment.** This paragraph applies to X-ray equipment for other than medical or dental use.

(1) **Disconnecting means.**

(i) A disconnecting means shall be provided in the supply circuit. The disconnecting means shall be operable from a location readily accessible from the X-ray control. For equipment connected to a 120-volt branch circuit of 30 amperes or less, a grounding-type attachment plug cap and receptacle of proper rating may serve as a disconnecting means.

(ii) If more than one piece of equipment is operated from the same high-voltage circuit, each piece or each group of equipment as a unit shall be provided with a high-voltage switch or equivalent disconnecting means. This disconnecting means shall be constructed, enclosed, or located so as to avoid contact by employees with its live parts.

(2) **Control.**

(i) **Radiographic and fluoroscopic types.** Radiographic and fluoroscopic-type equipment shall be effectively enclosed or shall have interlocks that de-energize the equipment automatically to prevent ready access to live current-carrying parts.

(ii) **Diffraction and irradiation types.** Diffraction-and irradiation-type equipment shall be provided with a means to indicate when it is energized unless the equipment or installation is effectively enclosed or is provided with a means to indicate when it is energized unless the equipment or installation is effectively enclosed or is provided with interlocks to prevent access to live current-carrying parts during operation.

(g) **Induction and dielectric heating equipment.**

(1) **Scope.** Paragraphs (g)(2) and (g)(3) of this section cover induction and dielectric heating equipment and accessories for industrial and scientific applications, but not for medical or dental applications or for appliances.

(2) **Guarding and grounding.**

(i) **Enclosures.** The converting apparatus (including the DC line) and high-frequency electric circuits (excluding the output circuits and remote-control circuits) shall be completely contained within enclosures of noncombustible material.

(ii) **Panel controls.** All panel controls shall be of dead-front construction.

(iii) **Access to internal equipment.** Where doors are used for access to voltages from 500 to 1000 volts AC or DC, either door locks or interlocks shall be provided. Where doors are used for access to voltages of over 1000 volts AC or DC, either mechanical lockouts with a disconnecting means to prevent access until voltage is removed from the cubicle, or both door interlocking and mechanical door locks, shall be provided.

(iv) **Warning labels.** "Danger" labels shall be attached on the equipment and shall be plainly visible even when doors are open or panels are removed from compartments containing voltages of over 250 volts AC or DC.

(v) **Work applicator shielding.** Protective cages or adequate shielding shall be used to guard work applicators other than induction heating coils. Induction heating coils shall be protected by insulation and/or refractory materials. Interlock switches shall be used on all hinged access doors, sliding panels, or other such means of access to the applicator. Interlock switches shall be connected in such a manner as to remove all power from the applicator when any one of the access doors or panels is open. Interlocks on access doors or panels are not required if the applicator is an induction heating coil at DC ground potential or operating at less than 150 volts AC.

(vi) **Disconnecting means.** A readily accessible disconnecting means shall be provided by which each unit of heating equipment can be isolated from its supply circuit.

(3) **Remote control.** If remote controls are used for applying power, a selector switch shall be provided and interlocked to provide power from only one control point at a time. Switches operated by foot pressure shall be provided with a shield over the contact button to avoid accidental closing of the switch.

(h) **Electrolytic cells.**

(1) **Scope.** These provisions for electrolytic cells apply to the installation of the electrical components and accessory equipment of electrolytic cells, electrolytic cell lines, and process power supply for the production of aluminum, cadmium, chlorine, copper, fluorine, hydrogen peroxide, magnesium, sodium, sodium chlorate, and zinc. Cells used as a source of electric energy and for electroplating processes and cells used for production of hydrogen are not covered by these provisions.

(2) **Definitions applicable to this paragraph.**

Cell line: An assembly of electrically interconnected electrolytic cells supplied by a source of direct-current power.

Cell line attachment and auxiliary equipment: Cell line attachments and auxiliary equipment include, but are not limited to the following: auxiliary tanks; process piping; duct work; structural supports; exposed cell line conductors; conduits and other raceways; pumps; positioning equipment and cell cutout or bypass electrical devices. Auxiliary equipment also includes tools, welding machines, crucibles, and other portable equipment used for operation and maintenance within the electrolytic cell line working zone. In the cell line working zone, auxiliary equipment includes the exposed conductive surfaces of ungrounded cranes and crane-mounted cell-servicing equipment.

Cell-line working zone: The cell line working zone is the space envelope wherein operation or maintenance is normally performed on or in the vicinity of exposed energized surfaces of cell lines or their attachments.

Electrolytic cells: A receptacle or vessel in which electrochemical reactions are caused by applying energy for the purpose of refining or producing usable materials.

(3) **Application.** Installations covered by paragraph (h) of this section shall comply with all applicable provisions of this subpart, except as follows:

(i) Overcurrent protection of electrolytic cell DC process power circuits need not comply with the requirements of 1910.304(e).

(ii) Equipment located or used within the cell line working zone or associated with the cell line DC power circuits need not comply with the provisions of 1910.304(f).

(iii) Electrolytic cells, cell line conductors, cell line attachments, and the wiring of auxiliary equipment and devices within the cell line working zone need not comply with the provisions of 1910.303, and 1910.304(b) and (c).

(4) **Disconnecting means.**

(i) If more than one DC cell line process power supply serves the same cell line, a disconnecting means shall be provided on the cell line circuit side of each power supply to disconnect it from the cell line circuit.

(ii) Removable links or removable conductors may be used as the disconnecting means.

(5) **Portable electric equipment.**

(i) The frames and enclosures of portable electric equipment used within the cell line working zone may not be grounded. However, these frames and enclosures may be grounded if the cell line circuit voltage does not exceed 200 volts DC or if the frames are guarded.

(ii) Ungrounded portable electric equipment shall be distinctively marked and may not be interchangeable with grounded portable electric equipment.

(6) **Power supply circuits and receptacles for portable electric equipment.**

(i) Circuits supplying power to ungrounded receptacles for hand-held cord- and plug-connected equipment shall be electrically isolated from any distribution system supplying areas other than the cell line working zone and shall be ungrounded. Power for these circuits shall be supplied through isolating transformers.

(ii) Receptacles and their mating plugs for ungrounded equipment may not have provision for a grounding conductor and shall be of a configuration which prevents their use for equipment required to be grounded.

(iii) Receptacles on circuits supplied by an isolating transformer with an ungrounded secondary shall have a distinctive configuration, shall be distinctively marked, and may not be used in any other location in the plant.

(7) **Fixed and portable electric equipment.**

(i) AC systems supplying fixed and portable electric equipment within the cell line working zone need not be grounded.

(ii) Exposed conductive surfaces, such as electric equipment housings, cabinets, boxes, motors, raceways, and the like that are within the cell line working zone need not be grounded.

(iii) Auxiliary electrical devices, such as motors, transducers, sensors, control devices, and alarms, mounted on an electrolytic cell or other energized surface, shall be connected by any of the following means:

(A) Multiconductor hard usage or extra hard usage flexible cord;

(B) Wire or cable in suitable raceways; or

(C) Exposed metal conduit, cable tray, armored cable, or similar metallic systems installed with insulating breaks such that they will not cause a potentially hazardous electrical condition.

(iv) Fixed electric equipment may be bonded to the energized conductive surfaces of the cell line, its attachments, or auxiliaries. If fixed electric equipment is mounted on an energized conductive surface, it shall be bonded to that surface.

FEDERAL REGULATIONS

(8) **Auxiliary nonelectric connections.** Auxiliary nonelectric connections, such as air hoses, water hoses, and the like, to an electrolytic cell, its attachments, or auxiliary equipment may not have continuous conductive reinforcing wire, armor, braids, and the like. Hoses shall be of a nonconductive material.

(9) **Cranes and hoists.**

(i) The conductive surfaces of cranes and hoists that enter the cell line working zone need not be grounded. The portion of an overhead crane or hoist which contacts an energized electrolytic cell or energized attachments shall be insulated from ground.

(ii) Remote crane or hoist controls which may introduce hazardous electrical conditions into the cell line working zone shall employ one or more of the following systems:

(A) Insulated and ungrounded control circuit;

(B) Nonconductive rope operator;

(C) Pendant pushbutton with nonconductive supporting means and having conductive surfaces or ungrounded exposed conductive surfaces; or

(D) Radio.

(i) **Electrically driven or controlled irrigation machines.** (See 1910.302(b)(3).)

(1) **Lightning protection.** If an electrically driven or controlled irrigation machine has a stationary point, a driven ground rod shall be connected to the machine at the stationary point for lightning protection.

(2) **Disconnecting means.** The main disconnecting means for a center pivot irrigation machine shall be located at the point of connection of electrical power to the machine and shall be readily accessible and capable of being locked in the open position. A disconnecting means shall be provided for each motor and controller.

(j) **Swimming pools, fountains, and similar installations.**

(1) **Scope.** Paragraphs (j)(2) through (j)(5) of this section apply to electric wiring for and equipment in or adjacent to all swimming, wading, therapeutic, and decorative pools and fountains, whether permanently installed or storable, and to

metallic auxiliary equipment, such as pumps, filters, and similar equipment. Therapeutic pools in health care facilities are exempt from these provisions.

(2) **Lighting and receptacles.**

(i) **Receptacles.** A single receptacle of the locking and grounding type that provides power for a permanently installed swimmng pool recirculating pump motor may be located not less than 5 feet from the inside walls of a pool. All other receptacles on the property shall be located at least 10 feet from the inside walls of a pool. Receptacles which are located within 15 feet of the inside walls of the pool shall be protected by ground-fault circuit interrupters.

NOTE: In determining these dimensions, the distance to be measured is the shortest path the supply cord of an appliance connected to the receptacle would follow without piercing a floor, wall, or ceiling of a building or other effective permanent barrier.

(ii) **Lighting fixtures and lighting outlets.**

(A) Unless they are 12 feet above the maximum water level, lighting fixtures and lighting outlets may not be installed over a pool or over the area extending 5 feet horizontally from the inside walls of a pool. However, a lighting fixture or lighting outlet which has been installed before April 16, 1981, may be located less than 5 feet measured horizontally from the inside walls of a pool if it is at least 5 feet above the surface of the maximum water level and shall be rigidly attached to the existing structure. It shall also be protected by a ground-fault circuit interrupter installed in the branch circuit supplying the fixture.

(B) Unless installed 5 feet above the maximum water level and rigidly attached to the structure adjacent to or enclosing the pool, lighting fixtures and lighting outlets installed in the area extending between 5 feet and 10 feet horizontally from the inside walls of a pool shall be protected by a ground-fault circuit interrupter.

(3) **Cord- and plug-connected equipment.** Flexible cords used with the following equipment may not exceed 3 feet in length and shall have a copper equipment grounding conductor with a grounding-type attachment plug.

FEDERAL REGULATIONS 169

(i) Cord- and plug-connected lighting fixtures installed within 16 feet of the water surface of permanently installed pools.

(ii) Other cord- and plug-connected, fixed or stationary equipment used with permanently installed pools.

(4) **Underwater equipment.**

(i) A ground-fault circuit interrupter shall be installed in the branch circuit supplying underwater fixtures operating at more than 15 volts. Equipment installed underwater shall be approved for the purpose.

(ii) No underwater lighting fixtures may be installed for operation at over 150 volts between conductors.

(5) **Fountains.** All electric equipment operating at more than 15 volts, including power supply cords, used with fountains shall be protected by ground-fault circuit interrupters. (See 1910.302(b)(3).)

1910.307 Hazardous (classified) locations.

(a) **Scope.** This section covers the requirements for electric equipment and wiring in locations which are classified depending on the properties of the flammable vapors, liquids or gases, or combustible dusts or fibers which may be present therein and the likelihood that a flammable or combustible concentration or quantity is present. Hazardous (classified) locations may be found in occupancies such as, but not limited to, the following: aircraft hangars, gasoline dispensing and service stations, bulk storage plants for gasoline or other volatile flammable liquids, paint-finishing process plants, health care facilities, agricultural or other facilities where excessive combustible dusts may be present, marinas, boat yards, and petroleum and chemical processing plants. Each room, section, or area shall be considered individually in determining its classification. These hazardous (classified) locations are assigned six designations as follows:

Class I, Division 1
Class I, Division 2
Class II, Division 1
Class II, Division 2

Class III, Division 1

Class III, Division 2

For definitions of these locations see 1910.399(a). All applicable requirements in this subpart shall apply to hazardous (classified) locations, unless modified by provisions of this section.

(b) **Electrical installations.** Equipment, wiring methods, and installation of equipment in hazardous (classified) locations shall be intrinsically safe, approved for the hazardous (classified) location, or safe or for the hazardous (classified) location. Requirements for each of these options are as follows:

(1) **Intrinsically safe.** Equipment and associated wiring approved as intrinsically safe shall be permitted in any hazardous (classified) location for which it is approved.

(2) **Approved for the hazardous (classified) location.**

(i) Equipment shall be approved not only for the class of location but also for the ignitable or combustible properties of the specific gas, vapor, dust, or fiber that will be present.

NOTE: NFPA 70, the National Electrical Code, lists or defines hazardous gases, vapors, and dusts by "Groups" characterized by their ignitible or combustible properties.

(ii) Equipment shall be marked to show the class, group, and operating temperature or temperature range, based on operation in a 40 degrees C ambient, for which it is approved. The temperature marking may not exceed the ignition temperature of the specific gas or vapor to be encountered. However, the following provisions modify this marking requirement for specific equipment:

(A) Equipment of the non-heat-producing type, such as junction boxes, conduit, and fittings, and equipment of the heat-producing type having a maximum temperature not more than 100 degrees C (212 degrees F) need not have a marked operating temperature or temperature range.

(B) Fixed lighting fixtures marked for use in Class I, Division 2 locations only, need not be marked to indicate the group.

FEDERAL REGULATIONS

(C) Fixed general-purpose equipment in Class I locations, other than lighting fixtures, which is acceptable for use in Class I, Division 2 locations need not be marked with the class, group, division, or operating temperature.

(D) Fixed dust-tight equipment, other than lighting fixtures, which is acceptable for use in Class II, Division 3 and Class III locations need not be marked with the class, group, division, or operating temperature.

(3) **Safe for the hazardous (classified) location.** Equipment which is safe for the location shall be of a type and design which the employer demonstrates will provide protection from the hazards arising from the combustibility and flammability of vapors, liquids, gases, dust, or fibers.

NOTE: The National Electrical Code, NFPA 70, contains guidelines for determining the type and design of equipment and installations which will meet this requirement. The guidelines of this document address electric wiring, equipment, and systems installed in hazardous (classified) locations and contain specific provisions for the following: wiring methods, wiring connections; conductor insulation, flexible cords, sealing and drainage, transformers, capacitors, switches, circuit breakers, fuses, motor controllers, receptacles, attachment plugs, meters, relays, instruments, resistors, generators, motors, lighting fixtures, storage battery charging equipment, electric cranes, electric hoists and similar equipment, utilization equipment, signaling systems, alarm systems, remote control systems, local loud speaker and communication systems, ventilation piping, live parts, lightning surge protection, and grounding. Compliance with these guidelines will constitute one means, but not the only means, of compliance with this paragraph.

(c) **Conduits.** All conduits shall be threaded and shall be made wrench-tight. Where it is impractical to make a threaded joint tight, a bonding jumper shall be utilized.

(d) **Equipment in Division 3 locations.** Equipment that has been approved for a Division 1 location may be installed in a Division 2 location of the same class and group. General purpose equipment or equipment in general-purpose enclosures may be installed in Division 2 locations if the equipment does not constitute a source of ignition under normal operating conditions.

1910.308 Special systems.

(a) **Systems over 600 volts, nominal.** Paragraphs (a)(1) through (4) of this section cover the general requirements for all circuits and equipment operated at over 600 volts.

(1) **Wiring methods for fixed installations.**

(i) Above-ground conductors shall be installed in rigid metal conduit, in intermediate metal conduit, in cable trays, in cablebus, in other suitable raceways, or as open runs of metal-clad cable suitable for the use and purpose. However, open runs of non-metallic-sheathed cable or of bare conductors or busbars may be installed in locations accessible only to qualified persons. Metallic shielding components, such as tapes, wires, or braids for conductors, shall be grounded. Open runs of insulated wires and cables having a bare lead sheath or a braided outer covering shall be supported in a manner designed to prevent physical damage to the braid or sheath.

(ii) Conductors emerging from the ground shall be enclosed in approved raceways. (See 1910.302(b)(3).)

(2) **Interrupting and isolated devices.**

(i) Circuit breaker installations located indoors shall consist of metal-mounted units. In locations accessible only to qualified personnel, open mounting of circuit breakers is permitted. A means of indicating the open and closed position of circuit breakers shall be provided.

(ii) Fused cutouts installed in building or transformer vaults shall be of a type approved for the purpose. They shall be readily accessible for fuse replacement.

(iii) A means shall be provided to isolate equipment completely for inspection and repairs. Isolating means which are not designed to interrupt the load current of the circuit shall be either interlocked with an approved circuit interrupter or provided with a sign warning against opening them under load.

(3) **Mobile and portable equipment.**

(i) **Power cable connections to mobile machines.** A metallic enclosure shall be provided on the mobile machine for

FEDERAL REGULATIONS

enclosing the terminals of the power cable. The enclosure shall include provisions for a solid connection for the ground wire(s) terminal to ground the machine frame effectively. The method of cable termination used shall prevent any strain or pull on the cable from stressing the electrical connections. The enclosure shall have provision for locking so only authorized qualified persons may open it and shall be marked with a sign warning of the presence of energized parts.

(ii) **Guarding live parts.** All energized switching and control parts shall be enclosed in effectively grounded metal cabinets or enclosures. Circuit breakers and protective equipment shall have the operating means projecting through the metal cabinet or enclosure so these units can be reset without locked doors being opened. Enclosures and metal cabinets shall be locked so that only authorized qualified persons have access and shall be marked with a sign warning of the presence of energized parts. Collector ring assemblies on revolving-type machines (shovels, draglines, etc.) shall be guarded.

(4) **Tunnel installation.**

(i) **Application.** The provisions of this paragraph apply to installation and use of high-voltage power distribution and utilization equipment which is portable and/or mobile, such as substations, trailers, cars, mobile shovels, draglines, hoists, drills, dredges, compressors, pumps, conveyors, and underground excavators.

(ii) **Conductors.** Conductors in tunnels shall be installed in one or more of the following:

(A) Metal conduit or other metal raceway,

(B) Type MC cable, or

(C) Other approved multiconductor cable.

Conductors shall also be located or guarded as to protect them from physical damage. Multiconductor portable cable may supply mobile equipment. An equipment grounding conductor shall be run with circuit conductors inside the metal raceway or inside the multiconductor cable jacket. The equipment grounding conductor may be insulated or bare.

(iii) **Guarding live parts.** Bare terminals of transformers, switches, motor controllers, and other equipment shall be enclosed to prevent accidental contact with energized parts. Enclosures for use in tunnels shall be drip-proof, weatherproof, or submersible as required by the environmental conditions.

(iv) **Disconnecting means.** A disconnecting means that simultaneously opens all ungrounded conductors shall be installed at each transformer or motor location.

(v) **Grounding and bonding.** All nonenergized metal parts of electric equipment and metal raceways and cable sheaths shall be effectively grounded and bonded to all metal pipes and rails at the portal and at intervals not exceeding 1000 feet throughout the tunnel.

(b) **Emergency power systems.**

(1) **Scope.** The provisions for emergency systems apply to circuits, systems, and equipment intended to supply power for illumination and special loads, in the event of failure of the normal supply.

(2) **Wiring methods.** Emergency circuit wiring shall be kept entirely independent of all other wiring and equipment and may not enter the same raceway cable, box, or cabinet or other wiring except either where common circuit elements suitable for the purpose are required, or for transferring power from the normal to the emergency source.

(3) **Emergency illumination.** Where emergency lighting is necessary, the system shall be so arranged that the failure of any individual lighting element, such as the burning out of a light bulb, cannot leave any space in total darkness.

(c) **Class 1, Class 2, and Class 3 remote control, signaling, and power-limited circuits.**

(1) **Classification.** Class 1, Class 2, or Class 3 remote control, signaling, or power-limited circuits are characterized by their usage and electrical power limitation which differentiates them from light and power circuits. These circuits are classified in accordance with their respective voltage and power limitations as summarized in paragraphs (c)(1)(i) through (c)(1)(iii) of this section.

(i) **Class 1 circuits.**

(A) A Class 1 power-limited circuit is supplied from a source having a rated output of not more than 30 volts and 1000 volt-amperes.

(B) A Class 1 remote control circuit or a Class 1 signaling circuit has a voltage which does not exceed 600 volts; however, the power output of the source need not be limited.

(ii) **Class 2 and Class 3 circuits.**

(A) Power for Class 2 and Class 3 circuits is limited either inherently (in which no overcurrent protection is required) or by a combination of a power source and overcurrent protection.

(B) The maximum circuit voltage is 150 volts AC or DC for a Class 2 inherently limited power source, and 100 volts AC or DC for a Class 3 inherently limited power source.

(C) The maximum circuit voltage is 30 volts AC and 60 volts DC for a Class 2 power source limited by overcurrent protection, and 150 volts AC or DC for a Class 3 power source limited by overcurrent protection.

(iii) The maximum circuit voltages in paragraphs (c)(1)(i) and (c)(1)(ii) or this section apply to sinusoidal AC or continuous DC power sources, and where wet contact occurence is not likely.

(2) **Marking.** A Class 2 or Class 3 power supply unit shall be durably marked where plainly visible to indicate the class of supply and its electrical rating. (See 1910.302(b)(3).)

(d) **Fire protective signaling systems.** (See 1910.302(b)(3).)

(1) **Classifications.** Fire protective signaling circuits shall be classified either as non-power limited or power limited.

(2) **Power sources.** The power sources for use with fire protective signaling circuits shall be either power limited or nonlimited as follows:

(i) The power supply of non-power-limited fire protective signaling circuits shall have an output voltage not in excess of 600 volts.

(ii) The power for power-limited fire protective signaling circuits shall be either inherently limited, in which no overcurrent protection is required, or limited by a combination of a power source and overcurrent protection.

(3) **Non-power-limited conductor location.** Non-power-limited fire protective signaling circuits and Class 1 circuits may occupy the same enclosure, cable, or raceway provided all conductors are insulated for maximum voltage of any conductor within the enclosure, cable, or raceway. Power supply and fire protective signaling circuit conductors are permitted in the same enclosure, cable, or raceway only if connected to the same equipment.

(4) **Power-limited conductor location.** Where open conductors are installed, power-limited fire protective signaling circuits shall be separated at least 2 inches from conductors of any light, power, Class 1, and non-power-limited fire protective signaling circuits unless a special and equally protective method of conductor separation is employed. Cables and conductors of two or more power-limited fire protective signaling circuits or Class 3 circuits are permitted in the same cable, enclosure, or raceway. Conductors of one or more Class 2 circuits are permitted within the same cable, enclosure, or raceway with conductors of power-limited fire protective signaling circuits provided that the insulation of Class 2 circuit conductors in the cable enclosure, or raceway is at least that needed for the power-limited fire protective signaling circuits.

(5) **Identification.** Fire protective signaling circuits shall be identified at terminal and junction locations in a manner which will prevent unintentional interference with the signaling circuit during testing and servicing. Power-limited fire protective signaling circuits shall be durably marked as such where plainly visible at terminations.

(e) **Communications systems.**

(1) **Scope.** These provisions for communication systems apply to such systems as central-station-connected and non-central-station-connected telephone circuits, radio and television receiving and transmitting equipment, including community antenna television and radio distribution systems, telegraph, district messenger, and outside wiring for fire and burglar alarm, and similar central station systems. These installations need not comply with the provision of 1910.303 through 1910.308(d), except 1910.304(c)(1) and 1910.307(b).

(2) Protective devices.

(i) Communication circuits so located as to be exposed to accidental contact with light or power conductors operating at over 300 volts shall have each circuit so exposed provided with a protector approved for the purpose.

(ii) Each conductor of a lead-in from an outdoor antenna shall be provided with an antenna discharge unit or other suitable means that will drain static charges from the antenna system.

(3) Conductor location.

(i) Outside of buildings.

(a) Receiving distribution lead-in or aerial-drop cables attached to buildings and lead-in conductors to radio transmitters shall be so installed as to avoid the possibility of accidental contact with electric light or power conductors.

(b) The clearance between lead-in conductors and any lightning protection conductors may not be less than 6 feet.

(ii) **On poles.** Where practicable, communication conductors on poles shall be located below the light or power conductors. Communications conductors may not be attached to a crossarm that carries light or power conductors.

(iii) **Inside of buildings.** Indoor antennas, lead-ins, and other communication conductors attached as open conductors to the inside of buildings shall be located at least 2 inches from conductors of any light or power or Class 1 circuits unless a special and equally protective method of conductor separation, approved for the purpose, is employed.

(4) Equipment location. Outdoor metal structures supporting antennas, as well as self-supporting antennas such as vertical rods or dipole structures, shall be located as far away from overhead conductors of electric light and power circuits of over 150 volts to ground as necessary to avoid the possibility of the antenna or structure falling into or making accidental contact with such circuits.

(5) Grounding.

(i) **Lead-in conductors.** If exposed to contact with electric light and power conductors, the metal sheath of aerial cables entering buildings shall be grounded or shall be interrupted close to the entrance to the building by an insulating joint or equivalent device. Where protective devices are used, they shall be grounded in an approved manner.

(ii) **Antenna structures.** Masts and metal structures supporting antennas shall be permanently and effectively grounded without splice or connection in the grounding conductor.

(iii) **Equipment enclosures.** Transmitters shall be enclosed in a metal frame or grill or separated from the operating space by a barrier, all metallic parts of which are effectively connected to ground. All external metal handles and controls accessible to the operating personnel shall be effectively grounded. Unpowered equipment and enclosures shall be considered grounded where connected to an attached coaxial cable with an effectively grounded metallic shield.

[46 FR 4056, Jan. 16, 1981; 46 FR 40185, Aug. 7, 1981]

SAFETY-RELATED WORK PRACTICES

1910.309-1910.330 [Reserved]

SAFETY-RELATED MAINTENANCE REQUIREMENTS

1910.331-1910.360 [Reserved]

SAFETY REQUIREMENTS FOR SPECIAL EQUIPMENT

1910.361-1910.380 [Reserved]

OTHER

1910-381-1910.398 [Reserved]

DEFINITIONS

1910.399 Definitions applicable to this subpart.

(a) **Definitions applicable to 1910.302 through 1910.330.**

(1) **Acceptable.** An installation or equipment is acceptable to the Assistant Secretary of Labor, and approved within the meaning of this Subpart S:

FEDERAL REGULATIONS

(i) If it is accepted, or certified, or listed, or labeled, or otherwise determined to be safe by a nationally recognized testing laboratory, such as, but not limited to, Underwriters' Laboratories, Inc., and Factory Mutual Engineering Corp.; or

(ii) With respect to an installation or equipment of a kind which no nationally recognized testing laboratory accepts, certifies, lists, labels, or determines to be safe, if it is inspected or tested by another Federal agency, or by a State, municipal, or other local authority responsible for enforcing occupational safety provisions of the National Electrical Code, and found in compliance with the provisions of the National Electrical Code as applied in this Subpart; or

(iii) With respect to custom-made equipment or related installations which are designed, fabricated for, and intended for use by a particular customer, if it is determined to be safe for its intended use by its manufacturer on the basis of test data which the employer keeps and makes available for inspection to the Assistant Secretary and his authorized representatives.

(2) **Accepted.** An installation is "accepted" if it has been inspected and found by a nationally recognized testing laboratory to conform to specified plans or to procedures of applicable code.

(3) **Accessible.** (As applied to wiring methods.) Capable of being removed or exposed without damaging the building structure or finish, or not permanently closed in by the structure or finish of the building. (See "**concealed**" and "**exposed.**")

(4) **Accessible.** (As applied to equipment.) Admitting close approach; not guarded by locked doors, elevation, or other effective means. (See "**Readily accessible.**")

(5) **Ampacity.** Current-carrying capacity of electric conductors expressed in amperes.

(6) **Appliances.** Utilization equipment, generally other than industrial, normally built in standardized sizes or types, which is installed or connected as a unit to perform one or more functions such as clothes washing, air conditioning, food mixing, deep frying, etc.

(7) **Approved.** Acceptable to the authority enforcing this subpart. The authority enforcing this subpart is the Assistant Secretary of Labor for Occupational Safety and Health. The definition of "acceptable" indicates what is acceptable to the Assistant Secretary of Labor, and therefore approved within the meaning of this Subpart.

(8) **Approved for the purpose.** Approved for a specific purpose, environment, or application described in a particular standard requirement.

Suitability of equipment or materials for a specific purpose, environment, or application may be determined by a nationally recognized testing laboratory, inspection agency, or other organization as part of its listing and labeling program. (See "Labeled" or "Listed.")

(9) **Armored cable.** Type AC armored cable is a fabricated assembly of insulated conductors in a flexible metallic enclosure.

(10) **Askarel.** A generic term for a group of nonflammable synthetic chlorinated hydrocarbons used as electrical insulating media. Askarels of various compositional types are used. Under arcing conditions, the gases produced, while consisting predominantly of noncombustible hydrogen chloride, can include varying amounts of combustible gases depending upon the askarel type.

(11) **Attachment plug (Plug cap) (Cap).** A device which, by insertion in a receptacle, establishes connection between the conductors of the attached flexible cord and the conductors connected permanently to the receptacle.

(12) **Automatic.** Self-acting, operating by its own mechanism when actuated by some impersonal influence, as, for example, a change in current strength, pressure, temperature, or mechanical configuration.

(13) **Bare conductor.** See "Conductor."

(14) **Bonding.** The permanent joining of metallic parts to form an electrically conductive path which will assure electrical continuity and the capacity to conduct safely any current likely to be imposed.

(15) **Bonding jumper.** A reliable conductor to assure the required electrical conductivity between metal parts required to be electrically connected.

(16) **Branch circuit.** The circuit conductors between the final overcurrent device protecting the circuit and the outlet(s).

(17) **Building.** A structure which stands alone or which is cut off from adjoining structures by fire walls with all openings therein protected by approved fire doors.

(18) **Cabinet.** An enclosure designed either for surface or flush mounting, and provided with a frame, mat, or trim in which a swinging door or doors are or may be hung.

(19) **Cable tray system.** A cable tray system is a unit or assembly of units or sections, and associated fittings, made of metal or other noncombustible materials forming a rigid structural system used to support cables. Cable tray systems include ladders, troughs, channels, solid bottom trays, and other similar structures.

(20) **Cablebus.** Cablebus is an approved assembly of insulated conductors with fittings and conductor terminations in a completely enclosed, ventilated, protective metal housing.

(21) **Center pivot irrigation machine.** A center pivot irrigation machine is a multi-motored irrigation machine which revolves around a central pivot and employs alignment switches or similar devices to control individual motors.

(22) **Certified.** Equipment is "certified" if it

(a) Has been tested and found by a nationally recognized testing laboratory to meet nationally recognized standards or to be safe for use in a specified manner, or

(b) Is of a kind whose production is periodically inspected by a nationally recognized testing laboratory, and

(c) It bears a label, tag, or other record of certification.

(23) **Circuit breaker.**

(i) (600 volts nominal, or less). A device designed to open and close a circuit by nonautomatic means and to open the circuit automatically on a predetermined overcurrent without injury to itself when properly applied within its rating.

(ii) (Over 600 volts, nominal). A switching device capable of making, carrying, and breaking currents under normal circuit conditions, and also making, carrying for a specified time, and breaking currents under specified abnormal circuit conditions, such as those of short circuit.

(24) **Class I locations.** Class I locations are those in which flammable gases or vapors are or may be present in the air in quantities sufficient to produce explosive or ignitible mixtures. Class I locations include the following:

(i) **Class I, Division 1.** A Class I, Division 1 location is a location

(a) In which hazardous concentrations of flammable gases or vapors may exist under normal operating conditions; or

(b) In which hazardous concentrations of such gases or vapors may exist frequently because of repair or maintenance operations or because of leakage; or

(c) In which breakdown or faulty operation of equipment or processes might release hazardous concentrations of flammable gases or vapors, and might also cause simultaneous failure of electric equipment.

NOTE: This classification usually includes locations where volatile flammable liquids or liquefied flammable gases are transferred from one container to another; interiors of spray booths and areas in the vicinity of spraying and painting operations where volatile flammable solvents are used; locations containing open tanks or vats of volatile flammable liquids; drying rooms or compartments for the evaporation of flammable solvents; locations containing fat and oil extraction equipment using volatile flammable solvents; portions of cleaning and dyeing plants where flammable liquids are used; gas generator rooms and other portions of gas manufacturing plants where flammable gas may escape; in adequately ventilated pump rooms for flammable gas or for volatile flammable liquids; the interiors of refrigerators and freezers in which volatile flammable materials are stored in open, lightly stoppered, or easily ruptured containers; and all other locations where ignitible concentrations of flammable vapors or gases are likely to occur in the course of normal operations.

(ii) **Class I, Division 2.** A Class I, Division 2 location is a location

(a) In which volatile flammable liquids or flammable gases are handled, processed, or used, but in which the hazardous liquids, vapors, or gases will normally be confined within closed containers or closed systems from which they can escape only in case of accidental rupture or breakdown of such containers or systems, or in case of abnormal operation of equipment; or

(b) In which hazardous concentrations of gases or vapors are normally prevented by positive mechanical ventilation, and which might become hazardous through failure or abnormal operations of the ventilating equipment; or

(c) That is adjacent to a Class I, Division 1 location, and to which hazardous concentrations of gases or vapors might occasionally be communicated unless such communication is prevented by adequate positive-pressure ventilation from a source of clean air, and effective safeguards against ventilation failure are provided.

NOTE: This classification usually includes locations where volatile flammable liquids or flammable gases or vapors are used, but which would become hazardous only in case of an accident or of some unusual operating condition. The quantity of flammable material that might escape in case of accident, the adequacy of ventilating equipment, the total area involved, and the record of the industry or business with respect to explosions or fires are all factors that merit consideration in determining the classification and extent of each location.

Piping without valves, checks, meters, and similar devices would not ordinarily introduce a hazardous condition even though used for flammable liquids or gases. Locations used for the storage of flammable liquids or a liquefied or compressed gases in sealed containers would not normally be considered hazardous unless also subject to other hazardous conditions.

Electrical conduits and their associated enclosures separated from process fluids by a single seal or barrier are classed as a Division 2 location if the outside of the conduit and enclosures is a nonhazardous location.

(25) **Class II locations.** Class II locations are those that are hazardous because of the presence of combustible dust. Class II locations include the following:

(i) **Class II, Division 1.** A Class II, Divison 1 location is a location:

(a) In which combustible dust is or may be in suspension in the air under normal operating conditions, in quantities sufficient to produce explosive or ignitible mixtures; or

(b) Where mechanical failure or abnormal operation of machinery or equipment might cause such explosive or ignitible mixtures to be produced, and might also provide a source of ignition through simultaneous failure of electric equipment, operation of protection devices, or from other causes, or

(c) In which combustible dusts of an electrically conductive nature may be present.

NOTE: This classification may include areas of grain handling and processing plants, starch plants, sugar-pulverizing plants, malting plants, hay-grinding plants, coal pulverizing plants, areas where metal dusts and powders are produced or processed, and other similar locations which contain dust producing machinery and equipment (except where the equipment is dust-tight or vented to the outside). These areas would have combustible dust in the air, under normal operating conditions, in quantities sufficient to produce explosive or ignitible mixtures. Combustible dusts which are electrically nonconductive include dusts produced in the handling and processing of grain products, pulverized sugar and cocoa, dried egg and milk powders, pulverized spices, starch and pastes, potato and woodflour, oil meal from beans and seed, dried hay, and other organic materials which may produce combustible dusts when processed or handled. Dusts containing magnesium or aluminum are particularly hazardous and the use of extreme caution is necessary to avoid ignition and explosion.

(ii) **Class II, Division 2.** A Class II, Division 2 location is a location in which

(a) Combustible dust will not normally be in suspension in the air in quantities sufficient to produce explosive or ignitible mixtures, and dust accumulations are normally insufficient to interfere with the normal operation of electrical equipment or other apparatus; or

(b) Dust may be in suspension in the air as a result of infrequent malfunctioning of handling or processing of equipment, and dust accumulations resulting therefrom may be ignitible by abnormal operation or failure of electrical equipment or other apparatus.

NOTE: This classification includes locations where dangerous concentrations of suspended dust would not be likely but where dust accumulations might form on or in the vicinity of electric equipment. These areas may contain equipment from which appreciable quantities of dust would escape under abnormal operating conditions or be adjacent to a Class II Division 1 location, as described above, into which an explosive or ignitible concentration of dust may be put into suspension under abnormal operating conditions.

(26) **Class III locations.** Class III locations are those that are hazardous because of the presence of easily ignitible fibers or flyings but in which such fibers or flyings are not likely to be in suspension in the air in quantities sufficient to produce ignitible mixtures. Class III locations include the following:

(i) **Class III, Division 1.** A Class III, Division 1 location is a location in which easily ignitible fibers or materials producing combustible flyings are handled, manufactured, or used.

NOTE: Such locations usually include some parts of rayon, cotton, and other textile mills; combustible fiber manufacturing and processing plants; cotton gins and cottonseed mills; flax-processing plants; clothing manufacturing plants; woodworking plants, and establishments; and industries involving similar hazardous processes or conditions.

Easily ignitible fibers and flyings include rayon, cotton (including cotton linters and cotton waste), sisal or henequen, istle, jute, hemp. tow, cocoa fiber, oakum, baled waste kapok, Spanish mose, excelsior, and other materials or similar nature.

(ii) **Class III, Division 2.** A Class III, Division 2 location is a location in which easily ignitible fibers are stored or handled, except in process of manufacture.

(27) **Collector ring.** A collector ring is an assembly of slip rings for transferring electrical energy from a stationary to a rotating member.

(28) **Concealed.** Rendered inaccessible by the structure or finish of the building. Wires in concealed raceways are considered consealed, even though they may become accessible by withdrawing them. [See "Accessible. **(As applied to wiring methods.)**"]

(29) **Conductor.**

(i) **Bare.** A conductor having no covering or electrical insulation whatsoever.

(ii) **Covered.** A conductor encased within material of composition or thickness that is not recognized as electrical insulation.

(iii) **Insulated.** A conductor encased within material of composition and thickness that is recognized as electrical insulation.

(30) **Conduit body.** A separate portion of a conduit or tubing system that provides access through a removable cover(s) to the interior of the system at a junction of two or more sections of the system or at a terminal point of the system. Boxes such as FS and FD or larger cast or sheet metal boxes are not classified as conduit bodies.

(31) **Controller.** A device or group of devices that serves to govern, in some predetermined manner, the electric power delivered to the apparatus to which it is connected.

(32) **Cooking unit, counter-mounted.** A cooking appliance designed for mounting in or on a counter and consisting of one or more heating elements, internal wiring, and built-in or separately mountable controls. (See "Oven, wall-mounted.")

(33) **Covered conductor.** See "Conductor."

(34) **Cutout.** (Over 600 volts, nominal.) An assembly of a fuse support with either a fuseholder, fuse carrier, or disconnecting blade. The fuseholder or fuse carrier may include a conducting

element (fuse link), or may act as the disconnecting blade by the inclusion of a nonfusible member.

(35) **Cutout box.** An enclosure designed for surface mounting and having swinging doors or covers secured directly to the telescoping with the walls of the box proper. (See "Cabinet.")

(36) **Damp location.** See "Location."

(37) **Dead front.** Without live parts exposed to a person on the operating side of the equipment.

(38) **Device.** A unit of an electrical system which is intended to carry but not utilize electric energy.

(39) **Dielectric heating.** Dielectric heating is the heating of a nominally insulating material due to its own dielectric losses when the material is placed in a varying electric field.

(40) **Disconnecting means.** A device, or group of devices, or other means by which the conductors of a circuit can be disconnected from their source of supply.

(41) **Disconnecting (or Isolating) switch.** (Over 600 volts, nominal.) A mechanical switching device used for isolating a circuit or equipment from a source of power.

(42) **Dry location.** See "Location."

(43) **Electric sign.** A fixed, stationary, or portable self-contained, electrically illuminated utilization equipment with words or symbols designed to convey information or attract attention.

(44) **Enclosed.** Surrounded by a case, housing, fence or walls which will prevent persons from accidentally contacting energized parts.

(45) **Enclosure.** The case of housing of apparatus, or the fence or walls surrounding an installation to prevent personnel from accidentally contacting energized parts, or to protect the equipment from physical damage.

(46) **Equipment.** A general term including material, fittings, devices, appliances, fixtures, apparatus, and the like, used as a part of, or in connection with, an electrical installation.

(47) **Equipment grounding conductor.** See "Grounding conductor, equipment."

(48) **Explosion-proof apparatus.** Apparatus enclosed in a case that is capable of withstanding an explosion of a specified gas or vapor which may occur within it and of preventing the ignition of a specified gas or vapor surrounding the enclosure by sparks, flashes, or explosion of the gas or vapor within, and which operates at such an external temperature that it will not ignite a surrounding flammable atmosphere.

(49) **Exposed.** (As applied to live parts.) Capable of being inadvertently touched or approached nearer than a safe distance by a person. It is applied to parts not suitably guarded, isolated, or insulated. (See "**Accessible.**" and "**Concealed.**")

(50) **Exposed.** (As applied to wiring methods.) On or attached to the surface or behind panels designed to allow access. [See "**Accessible.** (As applied to wiring methods.)"]

(51) **Exposed.** (For the purposes of 1910.308(e), **Communications systems.**) Where the circuit is in such a position that in case of failure of supports or insulation, contact with another circuit may result.

(52) **Externally operable.** Capable of being operated without exposing the operator to contact with live parts.

(53) **Feeder.** All circuit conductors between the service equipment, or the generator switchboard of an isolated plant, and the final branch-circuit overcurrent device.

(54) **Fitting.** An accessory such as a locknut, bushing, or other part of a wiring system that is intended primarily to perform a mechanical rather than an electrical function.

(55) **Fuse.** (Over 600 volts, nominal.) An overcurrent protective device with a circuit opening fusible part that is heated and severed by the passage of overcurrent through it. A fuse comprises all the parts that form a unit capable of performing the prescribed functions. It may or may not be the complete device necessary to connect it into an electrical circuit.

(56) **Ground.** A conducting connection, whether intentional or accidental, between an electrical circuit or equipment and the earth, or to some conducting body that serves in place of the earth.

(57) **Grounded.** Connected to earth or to some conducting body that serves in place of the earth.

(58) **Grounded, effectively.** (Over 600 volts, nominal.) Permanently connected to earth through a ground connection of sufficiently low impedance and having sufficient ampacity that ground fault current which may occur cannot build up to voltages dangerous to personnel.

(59) **Grounded conductor.** A system or circuit conductor that is intentionally grounded.

(60) **Grounding conductor.** A conductor used to connect equipment or the grounded circuit of a wiring system to a grounding electrode or electrodes.

(61) **Grounding conductor, equipment.** The conductor used to connect the non-current-carrying metal parts of equipment, raceways, and other enclosures to the system grounded conductor and/or the grounding electrode conductor at the service equipment or at the source of a separately derived system.

(62) **Grounding electrode conductor.** The conductor used to connect the grounding electrode to the equipment grounding conductor and/or to the grounded conductor of the circuit at the service equipment or at the source of a separately derived system.

(63) **Ground-fault circuit-interrupter.** A device whose function is to interrupt the electric circuit to the load when a fault current to ground exceeds some predetermined value that is less than that required to operate the overcurrent protective device of the supply circuit.

(64) **Guarded.** Covered, shielded, fenced, enclosed, or otherwise protected by means of suitable covers, casings, barriers, rails, screens, mats, or platforms to remove the likelihood of approach to a point of danger or contact by persons or objects.

(65) **Health care facilities.** Buildings or portions of buildings and mobile homes that contain, but are not limited to, hospitals, nursing homes, extended care facilities, clinics, and medical and dental offices, whether fixed or mobile.

(66) **Heating equipment.** For the purposes of 1910.306(g), the term "heating equipment" includes any equipment used for heating purposes if heat is generated by induction or dielectric methods.

(67) **Hoistway.** Any shaftway, hatchway, well hole, or other vertical opening or space in which an elevator or dumbwaiter is designed to operate.

(68) **Identified.** Identified, as used in reference to a conductor or its terminal, means that such conductor or terminal can be readily recognized as grounded.

(69) **Induction heating.** Induction heating is the heating of a nominally conductive material due to its own I^2R losses when the material is placed in a varying electromagnetic field.

(70) **Insulated conductor.** See "Conductor."

(71) **Interrupter switch.** (Over 600 volts, nominal.) A switch capable of making, carrying, and interrupting specified currents.

(72) **Irrigation machine.** An irrigation machine is an electrically driven or controlled machine, with one or more motors, not hand portable, and used primarily to transport and distribute water for agricultural purposes.

(73) **Isolated.** Not readily accessible to persons unless special means for access are used.

(74) **Isolated power system.** A system comprizing an isolating transformer or its equivalent, a line isolation monitor, and its ungrounded circuit conductors.

(75) **Labeled.** Equipment is "labeled" if there is attached to it a label, symbol, or other identifying mark of a nationally recognized testing laboratory which,

(a) Makes periodic inspections of the production of such equipment, and

(b) Whose labeling indicates compliance with nationally recognized standards or tests to determine safe use in a specified manner.

(76) **Lighting outlet.** An outlet intended for the direct connection of a lampholder, a lighting fixture, or a pendant cord terminating in a lampholder.

(77) **Listed.** Equipment is "listed" if it is of a kind mentioned in a list which,

(a) Is published by a nationally recognized laboratory which make periodic inspection of the production of such equipment, and

(b) States such equipment meets nationally recognized standards or has been tested and found safe for use in a specified manner.

(78) **Location.**

(i) **Damp location.** Partially protected locations under canopies, marquees, roofed open porches, and like locations, and interior locations subject to moderate degrees of moisture, such as some basements, some barns, and some cold-storage warehouses.

(ii) **Dry location.** A location not normally subject to dampness or wetness. A location classified as dry may be temporarily subject to dampness or wetness, as in the case of a building under construction.

(iii) **Wet location.** Installations underground or in concrete slabs or masonry in direct contact with the earth, and locations subject to saturation with water or other liquids, such as vehicle-washing areas, and locations exposed to whether and unprotected.

(79) **Medium voltage cable.** Type MV medium voltage cable is a single or multiconductor solid dielectric insulated cable rated 2000 volts or higher.

(80) **Metal-clad cable.** Type MC cable is a factory assembly of one or more conductors, each individually insulated and enclosed in a metallic sheath of interlocking tape, or a smooth or corrugated tube.

(81) **Mineral-insulated metal-sheathed cable.** Type MI mineral-insulated metal-sheathed cable is a factory assembly of one or more conductors insulated with a highly compressed refractory mineral insulation and enclosed in a liquidtight and gastight continuous copper sheath.

(82) **Mobile X-ray.** X-ray equipment mounted on a permanent base with wheels and/or casters for moving while completely assembled.

(83) **Nonmetallic-sheathed cable.** Nonmetallic-sheathed cable is a factory assembly of two or more insulated conductors

having an outer sheath of moisture resistant, flame-retardant, nonmetallic material. Nonmetallic sheathed cable is manufactured in the following types;

(i) **Type NM.** The overall covering has a flame-retardant and moisture-resistant finish.

(ii) **Type NMC.** The overall covering is flame-retardant, moisture-resistant, fungus-resistant, and corrosion-resistant.

(84) **Oil (filled) cutout.** (Over 600 volts, nominal.) A cutout in which all or part of the fuse support and its fuse link or disconnecting blade are mounted in oil with complete immersion of the contacts and the fusible portion of the conducting element (fuse link), so that arc interruption by severing of the fuse link or by opening of the contacts will occur under oil.

(85) **Open wiring on insulators.** Open wiring on insulators is an exposed wiring method using cleats, knobs, tubes, and flexible tubing for the protection and support of single insulated conductors run in or on buildings, and not concealed by the building structure.

(86) **Outlet.** A point on the wiring system at which current is taken to supply utilization equipment.

(87) **Outline lighting.** An arrangement of incandescent lamps or electric discharge tubing to outline or call attention to certain features such as the shape of a building or the decoration of a window.

(88) **Oven, wall-mounted.** An oven for cooking purposes designed for mounting in or on a wall or other surface and consisting of one or more heating elements, internal wiring, and built-in or separately mountable controls. (See "**Cooking unit, counter-mounted.**")

(89) **Overcurrent.** Any current in excess of the rated current of equipment or the ampacity of a conductor. It may result from overload (see definition), short circuit, or ground fault. A current in excess of rating may be accomodated by certain equipment and conductors for a given set of conditions. Hence, the rules for overcurrent protection are specific for particular situations.

(90) **Overload.** Operation of equipment in excess of normal, full load rating, or of a conductor in excess of rated ampacity which, when it persists for a sufficient length of time, would cause damage or dangerous overheating. A fault, such as a

short circuit or ground fault, is not an overload. (See "Overcurrent.")

(91) **Panelboard.** A single panel or group of panel units designed for assembly in the form of a single panel; including buses, automatic overcurrent devices, and with or without switches for the control of light, heat, or power circuits; designed to be placed in a cabinet or cutout box placed in or against a wall or partition and accessible only from the front. (See "Switchboard.")

(92) **Permanently installed decorative fountains and reflection pools.** Those that are constructed in the ground, on the ground, or in a building in such a manner that the pool cannot be readily disassembled for storage and are served by electrical circuits of any nature. These units are primarily constructed for their aesthetic value and not intended for swimming or wading.

(93) **Permanently installed swimming pools, wading and therapeutic pools.** Those that are constructed in the ground, on the ground, or in a building in such a manner that the pool cannot be readily disassembled for stroage whether or not served by electrical circuits of any nature.

(94) **Portable X-ray.** X-ray equipment designed to be hand-carried.

(95) **Power and control tray cable.** Type TC power and control tray cable is a factory assembly of two or more insulated conductors, with or without associated bare or covered grounding conductors under nonmetallic sheath, approved for installation in cable trays, in raceways, or where supported by a messenger wire.

(96) **Power fuse.** (Over 600 volts, nominal.) See "Fuse".

(97) **Power-limited tray cable.** Type PLTC nonmetallic-sheathed power limited tray cable is a factory assembly of two or more insulated conductors under a nonmetallic jacket.

(98) **Power outlet.** An enclosed assembly which may include receptacles, circuit breakers, fuseholders, fused switches, buses, and watt-hour meter mounting means; intended to supply and control power to mobile homes, recreational vehicles or boats, or to serve as a means for distributing power required to operate mobile or temporaily installed equipment.

(99) **Premises wiring system.** That interior and exterior wiring, including power, lighting, control, and signal circuit wiring together with all of its associated hardware, fittings, and wiring devices, both permanently and temorarily installed, which extends from the load end of the service drop, or load end of the service lateral conductors to the outlet(s). Such wiring does not include wiring internal to appliances, fixtures, motors, controllers, motor control centers, and similar equipment.

(100) **Qualified person.** One familiar with the construction and operation of the equipment and the hazards involved.

(101) **Raceway.** A channel designed expressly for holding wires, cables, or busbars, with additional functions as permitted in this subpart. Raceways may be of metal or insulating material, and the term includes rigid metal conduit, rigid nonmetallic conduit, intermediate metal conduit, liquidtight flexible metal conduit, flexible metallic tubing, flexible metal conduit, electrical metallic tubing, underfloor raceways, cellular concrete floor raceways, cellular metal floor raceways, surface raceways, wireways, and busways.

(102) **Readily accessible.** Capable of being reached quickly for operation, renewal, or inspections, without requiring those to whom ready access is requisite to climb over or remove obstacles or to resort to portable ladders, chairs, etc. (See "Accessible.")

(103) **Receptacle.** A receptacle is a contact device installed at the outlet for the connection of a single attachment plug. A single receptacle is a single contact device with no other contact device on the same yoke. A multiple receptacle is a single device containing two or more receptacles.

(104) **Receptacle outlet.** An outlet where one or more receptacles are installed.

(105) **Remote-control circuit.** Any electric circuit that controls any other circuit through a relay or an equivalent device.

(106) **Sealable equipment.** Equipment enclosed in a case or cabinet that is provided with a means of sealing or locking so that live parts cannot be made accessible without opening the enclosure. The equipment may or may not be operable without opening the enclosure.

(107) **Separately derived system.** A premises wiring system whose power is derived from generator, transformer, or converter winding and has no direct electrical connection, including a solidly connected grounded circuit conductor, to supply conductors originating in another system.

(108) **Service.** The conductors and equipment for delivering energy from the electricity supply system to the wiring system of the premises served.

(109) **Service cable.** Service conductors made up in the form of a cable.

(110) **Service conductors.** The supply conductors that extend from the street main or from transformers to the service equipment of the premises supplied.

(111) **Service drop.** The overhead service conductors from the last pole or other aerial support to and including the splices, if any, connecting to the service-entrance conductors at the building or other structure.

(112) **Service-entrance cable.** Service-entrance cable is a single conductor or multiconductor assembly provided with or without an overall covering, primarily used for services and of the following types:

(i) **Type SE,** having a flame-retardant, moisture-resistant covering, but not required to have inherent protection against mechanical abuse.

(ii) **Type USE,** recognized for underground use, having a moisture-resistant covering, but not required to have a flame-retardant covering or inherent protection against mechanical abuse. Single-conductor cables having an insulation specifically approved for the purpose do not require an outer covering.

(113) **Service-entrance conductors, overhead system.** The service conductors between the terminals of the service equipment and a point usually outside the building, clear of building walls, where joined by tap or splice to the service drop.

(114) **Service entrance conductors, underground system.** The service conductors between the terminals of the service equipment and the point of connection to the service lateral. Where service equipment is located outside the building

walls, there may be no service-entrance conductors, or they may be entirely outside the building.

(115) **Service equipment.** The necessary equipment, usually consisting of a circuit breaker or switch and fuses, and their accessories, located near the point of entrance of supply conductors to a building or other structure, or an otherwise defined area, and intended to constitute the main control and means of cutoff of the supply.

(116) **Service raceway.** The raceway that encloses the service-entrance conductors.

(117) **Shielded nonmetallic-sheathed cable.** Type SNM, shielded nonmetallic-sheathed cable is a factory assembly of two or more insulated conductors in an extruded core of moisture-resistant, flame-resistant nonmetallic material, covered with an overlapping spiral metal tape and wire shield and jacketed with an extruded moisture-, flame-, oil-, corrosion-, fungus-, and sunlight-resistant nonmetallic material.

(118) **Show window.** Any window used or designed to be used for the display of goods or advertising material, whether it is fully or partly enclosed or entirely open at the rear and whether or not it has a platform raised higher than the street floor level.

(119) **Sign.** See "Electric Sign."

(120) **Signaling circuit.** Any electric circuit that energizes signaling equipment.

(121) **Special permission.** The written consent of the authority having jurisdiction.

(122) **Storable swimming or wading pool.** A pool with a maximum dimension of 15 feet and a maximum wall height of 3 feet and is so constructed that it mat be readily disassembled for storage and reassembled to its original integrity.

(123) **Switchboard.** A large single panel, frame, or assembly of panels which have switches, buses, instruments, overcurrent and other protective devices generally accessible from the rear as well as from the front and are not intended to be installed in cabinets. (See "**Panelboard.**")

(124) **Switches.**

(i) **General-use switch.** A switch intended for use in general distribution and branch circuits. It is rated in amperes, and it is capable of interrupting its rated current at its rated voltage.

(ii) **General-use snap switch.** A form of general-use switch so constructed that it can be installed in flush device boxes or on outlet box covers, or otherwise used in conjunction with wiring systems recognized by this subpart.

(iii) **Isolating switch.** A switch intended for isolating an electric circuit from the source of power. It has no interrupting rating, and it is intended to be operated only after the circuit has been opened by some other means.

(iv) **Motor-circuit switch.** A switch, rated in horse-power, capable of interrupting the maximum operating overload current of a motor of the same horsepower rating as the switch at the rated voltage.

(125) **Switching devices.** (Over 600 volts, nominal.) Devices designed to close and/or open one or more electric circuits. Included in this category are circuit breakers, cutouts, disconnecting (or isolating) switches, disconnecting means, interrupter switches, and oil (filled) cutouts.

(126) **Transportable X-ray.** X-ray equipment installed in a vehicle or that may readily be disassembled for transport in a vehicle.

(127) **Utilization equipment.** Utilization equipment means equipment which utilizes electric energy for mechanical, chemical, heating, lighting, or similar useful purpose.

(128) **Utilization system.** A utilization system is a system which provides electric power and light for employee workplaces, and includes the premises wiring system and utilization equipment.

(129) **Ventilated.** Provided with a means to permit circulation of air sufficient to remove an excess of heat, fumes, or vapors.

(130) **Volatile flammable liquid.** A flammable liquid having a flash point below 38 degrees C (100 degrees F) or whose temperature is above its flash point.

(131) **Voltage (of a circuit).** The greatest root-mean-square (effective) difference of potential between any two conductors of the circuit concerned.

(132) **Voltage, nominal.** A nominal value assigned to a circuit or system for the purpose of conveniently designating its voltage class (as 120/240, 480Y/277, 600, etc.). The actual voltage at which a circuit operates can vary from the nominal within a range that permits satisfactory operation of equipment.

(133) **Voltage to ground.** For grounded circuits, the voltage between the given conductor and that point or conductor of the circuit that is grounded; for ungrounded circuits, the greatest voltage between the given conductor and any other conductor of the circuit.

(134) **Watertight.** So constructed that moisture will not enter the enclosure.

(135) **Weatherproof.** So constructed or protected that exposure to the weather will not interfere with successful operation. Rainproof, raintight, or watertight equipment can fulfill the requirements for weatherproof where varying weather conditions other than wetness, such as snow, ice, dust, or temperature extremes, are not a factor.

(136) **Wet location.** See "Location."

(137) **Wireways.** Wireways are sheetmetal troughs with hinged or removable covers for housing and protecting electric wires and cable and in which conductors are laid in place after the wireway has been installed as a complete system.

(b) **Definitions applicable to 1910.331 through 1910.360** [Reserved].

(c) **Definitions applicable to 1910.361 through 1910.380** [Reserved].

(d) **Definitions applicable to 1910.381 through 1910.398** [Reserved].

7
Case Histories and Electrical Workers Survey

CASE HISTORIES

Electrical Distribution Accidents

According to "Accident Facts," 1986 edition, published by the National Safety Council, in 1983 (latest available information) electric current killed 872 people, of whom 158 were electrocuted in the power generating and transmitting industry. In addition to deaths, the 1985 record shows an accident rate of 4.72 per 100 employees for the utilities industry.

Typical of accidents occurring with high voltage transmission power lines on transformer platforms is the recent accident in Los Alamos, New Mexico. Two teenage boys decided to climb up a facility supporting a step-down transformer in a school yard. In this case, there was no lock-out or barrier system to discourage this activity. However, the platform was about ten feet off the ground and danger signs were posted. According to articles in the *Los Alamos Monitor*, one of the boys' heads touched a bare 7,500-volt wire. The shock he received caused him to fall from the platform, from where he suffered a concussion. The school and the county utility department were sued for damages.

Typical accidents involving tree-trimming procedures and other elevated operations are the touching of power lines with conducting tools or equipment held by the craftsmen or operators that are grounded and who are not wearing insulating gloves, or who may not even be aware of the danger of being close to transmitting overhead lines.

Research Facility Accidents

Typical of accidents involving apparatus used in myriad experimental setups are four accidents associated with National Laboratories.

1. At the Lawrence Berkeley Laboratory, a laser physicist entered a laser laboratory after hours to tour acquaintances when he was electrocuted by a live circuit unprotected by the usual barriers and restraints.

2. At the Los Alamos National Laboratory, an electronics technician was troubleshooting a live power supply in a laser system when his elbow contacted a metal table top as his hand touched an energized electrical source, resulting in burns at the touched areas and in mental shock (three days lost time). Suspecting trouble, the experienced technician had placed one hand behind his back. No warning light had been assigned to the power supply to indicate when it was energized.

3. Also at Los Alamos, an electronic technician, recently hired, had not been checked out on the procedure for salvaging small capacitors (clystrons). Instead of deadshorting the terminals, or measuring voltage with a meter, or wearing protective gloves, he proceeded to lift the unit with his bare hand. He has scars on his fingertips and knuckles on one hand.

4. An electrocution occurred at Los Alamos when a manufacturer's representative was providing service to the electrical system of an industry x-ray machine. Although a safety watch was present during work on energized electrical circuits as required by Laboratory policy, the troubleshooter was operating in very limited work space, and his head contacted a live circuit as his arm pressed against a grounded support. The path of electron flow through his body and the amperage was such that he died instantly. CPR was administered immediately, but revival was not possible.

Other case histories could be described, but all electrical accidents are caused for the traditional reasons: unsafe conditions or/and unsafe actions. A good movie was made on electrical safety by the Safety Department, Lawrence Berkeley Laboratory of the University of California, and is recommended for indoctrination and training. It is titled, "High Voltage Angel" and may be obtained on a short-term loan basis.

ELECTRICAL WORKERS SURVEY

An effective electrical hazard control program is essential in reducing the frequency of such accidents described above. One approach to implementing such a program, or to evaluate an ongoing electrical safety program is to survey the entity involved. For example, the Los Alamos National Laboratory conducted such a survey, and the details are included herein as a guide that could be used.

Report of the Electrical Safety Review Committe of the Los Alamos National Laboratory

I. INTRODUCTION

The Electrical Safety Review Committee was appointed by the Director's Office on February 1, 1984, to examine the policies, procedures, and practices pertaining to electrical safety at the Laboratory. The adequacy of the Laboratory's electrical safety system was to be evaluated and recommendations for improvement made. Some specific questions to be addressed are set forth in the Committee Charter. This report presents the findings and recommendations of the Committee.

II. OBJECTIVE

The ultimate objective of the Committee is to contribute to the continuing effort of making the Laboratory a safer workplace. The Charter of the Committee states "The Electrical Safety Review Committee will conduct a review of electrical safety within the Laboratory," and then defines the functional and administrative areas to be reviewed.

In fulfilling this directive, the Committee conducted a broad survey of electrical work by means of questionnaires and interviews, and the results are set forth in Section V, GENERAL FINDINGS, and Section VI, RECOMMENDATIONS. We believe the Committee has met the Charter objectives and has generated a data base that can be used for future electrical safety assessments.

III. INVESTIGATIVE APPROACH

The scope and diversity of electrical work at the Laboratory is truely impressive. Power, control, instrumentation, or experimental systems pervade the work of nearly every technical group. The work is performed in facilities that vary in age and condition and by personnel varying in electrical skills. Much equipment is experimental in nature, requiring more or less continuous maintenance and modification. In the experience of the Committee members, individual and organizational attitudes and practices with respect to electrical safety are nearly as diverse as the work itself.

The Committee viewed electrical safety as a Laboratorywide responsibility and avoided specific criticism of divisions, groups, or individuals. The body of the report discusses our findings in this fashion.

The Committee was interested in both worker's and administrator's views. For this reason, these groups were polled separately.

Considering the scope of the subject, the Committee decided early in its deliberations to seek input from individuals throughout the Laboratory who are directly engaged in, or responsible for, electrical work and from those who may work in an environment in which electrical equipment constitutes a potentially dominant hazard. We believe that the opinions, advice, and criticism from such people are essential for a proper evaluation of our electrical safety system. Input was obtained by mailout questionnaires and by personal interviews. The interviews were used to explore viewpoints that might not be expressed on the questionnaires.

Two questionnaires were prepared: one for technicians and nonmanagement staff members and one for managers. In creating the questionnaires, several questions had to be addressed:

- What questions should be asked?
- How should the questions be stated?

- Who should be asked?
- Can the answers be statistically analyzed?

Experienced help was obtained in analyzing the questions. Because the Committee was seeking frank answers, the manner in which some questions were asked was important. The responses were anonymous, but divisions were identified.

The Committee selected 75 Laboratory organizations for the survey. Although the list is not all inclusive, the 75 organizations were identified where electrical safety concerns were of some significance:

Accelerator Division,
    Groups 1, 2, 3, 4, 5, 7
Chemistry Division,
    Groups 3, 4, 5, 6
Controlled Thermonuclear Division,
    Groups 2, 3, 4, 5, 8, 9
Earth Sciences Division,
    Groups 1, 6, 11
Isotope Nuclear Chemistry,
    Groups 3, 4, 5, 7, 11
Life Sciences Division
    Groups 1, 3, 4
Dynamic Test Division,
    Groups 1, 3, 4, 6, 7
Meson Physics Division,
    Groups 1, 2, 4, 7, 8, 11, 13

Materials Science Technology,
    Groups 3, 5, 6, 9, 11, 12, 13, 14
Physics Division,
    Groups 1, 2, 3, 4, 5, 7, 8, 9, 10, 11, 12, 14, 15, 16
Reactor Physics Division,
    Groups 1, 2, 8, 13
Experimental Weapons Division,
    Groups 5, 10, 11, 12
Facilities Engineering Division
Zia Company
    (Crafts Contractor)

Rosters were obtained of these groups for three categories: technicians, nonmanagement staff members, and management-series personnel. We initially considered several sampling schemes for sending out questionnaires. However, the problems of obtaining representative samples from diverse groups of varying sizes made it evident that, for a modest increase in the mailing, nearly all the personnel of interest could be covered. After eliminating nontechnical categories, the Committee distributed questionnaires to approximately 800 technicians, 600 nonmanagement staff members, and 250 managers. Experence indicates that responses to questionnaires normally run between 25 and 35% of the mailing. Because of the subject matter and R. N. Thorn's two memos, the response ran considerably higher. We received 874 responses out of the total mailing of approximately 1650. Returns from managers ran 47% and from Staff Member/Technicians (SM/TEC) 55%.

The Committee members also performed as many interviews as time allowed. We did not use a formal sampling scheme for the interviews. Individuals were selected on the basis of working in particularly significant (electrically) areas, having key responsibilities, or because they were known by the Committee to hold strong opinions. On the cover memo to the questionnaires, the Committee also invited anyone to contact it if they desired an interview. Seventy interviews were conducted.

IV. QUESTIONNAIRE DATA

A. In this section, the results of the SM/TEC and Management Questionnaires are presented in abbreviated form. The first four questions on the SM/TEC questionnaire were used to establish the type of work the respondent performs. This enabled the Committee to distinguish "electrical workers" (i.e., those who regularly worked with potentially hazardous voltages) from others. The questionnaires also enabled the Committee to distinguish between those who worked with intermediate voltages (40-600V) and those exposed to high voltages (over 600V). Most of the Committee's conclusions concerning the questionnaires are based on the responses of all electrical workers.

The results of the SM/TEC questionnaires were tabulated for each question both by division and by Laboratory total. The findings and recommendations of the Committee are generally based on Laboratory electrical worker totals. It is difficult (and in many instances improper) to compare division responses because of the widely varying circumstances within each organization. However, we felt that the data would be useful for division managers to examine for themselves.

In addition to the question responses, a large number of written comments were received including several lengthy essays. These are of considerable importance and the comments were categorized and tabulated. Also included in this tabulation are the comments received during the course of 70 interviews with personnel throughout the Laboratory.

B. SM/TEC Questionnaire Results. (The questions are not included, but the results are self-explanatory.)

Q-5. On CPR courses: of electrical workers, (unless otherwise differentiated, "electrical workers" means both the 40-600V and the over 600V respondents) 30% have never had a CPR course and 20% have had one more than three years ago. Therefore, 50% of all electrical workers are not current on CPR.

Q-6. On Health, Safety and Environment Division's electrical safety course: 60% of the 40-600V workers and 45% of the over 600V workers have not had an HSE electrical safety course.

Q-7 and 8. On shocks and burns: 21.4% of electrical workers reported one or more shocks in the past year (163 people). 1.3% reported electrical burns (10 people).

Q-9. On reporting incidents: 15.2% of electrical workers believe that electrical incidents are reported 1/3 of the time or less. (For the 40-600V workers the percentage was 13.5%, and for the over 600V workers it was 42.5%.)

Q-10. On causes of electrical accidents: The ranking for all electrical workers was

1. Rushed or distracted personnel.
2. Failure to verify that equipment was turned off.
3. Inadequate training.

Q-11, 12, 13, 14. On written Standard Operating Procedures (SOPs): A majority of electrical workers believe SOPs are effective in preventing accidents. More workers follow the "spirit" of an SOP rather than the letter. Only a minority of respondents always use SOPs for work involving hazards.

Q-15. On Group attitude: Of electrical workers, 64% believe that electrical safety is emphasized in their group. Of the remaining 36%, most feel it is left to the individual.

Strong statistical relationships were found to exist between responses concerning the groups' attitude toward electrical safety and those on their implementation of Laboratory electrical policy. (There is no statistical dependency existing between the sections'/groups' attitude toward electrical safety and the number of shocks that their personnel receive. This lack of dependency is attributed to the intervention of other factors, such as the amount of electrical work done by the personnel.) The population for these tests was composed of Staff Members (SMs) and Technicians (TECs) who had worked with equipment in the 40-600 volts range.

On organizations where respondents indicated that electrical safety was not emphasized:

- Personnel are less likely to have taken the HSE Division Safety course within the last year.
- Personnel are less likely to follow SOPs/SWPs (Special Work Permits).
- Personnel are less inclined to believe that following SOPs/SWPs can prevent accidents.
- Knowledge of HSE Divisions Manual Number AR7-1 on electrical safety requirements is less likely to be required of personnel.
- Personnel are less likely to answer that their electrical equipment is analyzed for electrical safety.
- Personnel are less likely to receive information on modifications to electrical equipment.
- Personnel are more likely to work alone on energized, exposed electrical equipment.

Q-16. On AR7-1: 53.2% of electrical workers are not familiar with the requirements of Section 7-1 in the HSE manual.

Q-17 & 18. On electrical equipment: 57.2% of all electrical workers responded that their group does not analyze electrical equipment for safety, or does so only occasionally.

Q-19. On working with electrical equipment: 36.2% of electrical workers who had an opinion believe that personnel in their group work alone on exposed energized electrical equipment 50% of the time or more.

Q-20. On reasons for working alone: Electrical workers ranked reasons 1 (most important) to 3: (Numbers are averaged rankings)

1.4 Did not think it was necessary.

1.8 Could not get second person.

2.4 Group/Section puts other concerns first.

Q-21. On lock-out tagging: 35% of electrical workers responded that their section does not use lock-out and tagging procedures. 45% of the over 600V workers do not.

Q-22. On electrical power problems: 23% of all electrical workers have experienced problems with power systems that did or could have caused an accident.

Q-23. On causes of power system problems: Ranking by the 23% in Q-22 was:

1. Improper wiring of plugs and outlets (25%) and lack of standards for power connectors (25%).

2. Inadequate drawings and schematics (19%).

3. Poor inspection of power systems (15%) and inadequate maintenance (15%).

C. Management Questionnaire Results

MQ-1 & 2. On Electrical concerns: 50% of managers believe that electric shock and burn hazards are a major concern in their areas of responsibility. 63% believe they are a major concern at the Laboratory.

MQ-3. On the three most significant potential electrical hazards in their group or division:

1. Power circuits over 100V.
2. Power supplies.
3. Capacitors.
4. Instrumentation and control systems.
5. Pulsed power.
6. Electrical conductors and connectors.
7. Power tools.

MQ-4. On adequacy of Laboratory's overall electrical safety program:

67% Adequate.

25% Generally adequate, but does not appear to address specific hazards of their operations.

8% Inadequate and needs improvement.

MQ-5. On three most prevalent hazards in area of responsibility:
1. Machinery and other.
2. Electrical.
3. Lasers.
4. Radioactive materials.
5. Explosives.
6. Cryogenics.
7. Lifting and moving.

MQ-6. On a safety policy specifically addressing the electrical hazards in their area of responsibility:

74.3% of managers say they have one.

25.7% say they do not.

MQ-7. On the ways in which their organization addresses electrical safety issues: (most respondents checked more than one)

100% Safety officer.
93% Safety meetings.
77% Electrical safety inspections.
76% SOPs and SWPs.
62% Informal discussion.
32% Accident analysis.
20% Equipment analysis.
8% Other.

MQ-8. On whether work places have been formally evaluated for electrical hazards.

51% yes.
49% no.

MQ-9. On reviewing electrical SOPs: (some checked more than one.

13% Not needed.
35% At least annually.
48% When changes are made to equipment.
19% When reminded.
14% Never.

MQ-10. On circumstances that may require compromise with the Laboratory's electrical safety requirements: Managers ranked reasons 1 (most important to 4: (Numbers are averaged rankings)

1.5 Other (Policy is inflexible, inconsistent, or irrelevant (4), employee lack of interest in safety (4), identifying problem areas (3), remote field areas (3), our safety is not compromised (12).

1.7 Schedule constraints

2.0 Insufficient budget

2.4 Insufficient emphasis by upper management

MQ-11, 12, & 15. On modifying electrical equipment:

81% of managers' organizations modify electrical equipment.

24% Always document modifications.

56% Say they have a communications system for informing personnel about critical electrical design information.

MQ-13 & 14. On Administrative Requirement AR7-1:

40% do not routinely apply AR7-1 in their work.

51% have not distributed AR7-1 or discussed with their personnel.

MQ-16. On whether they believe their personnel follow electrical safety policies:

17.4% Always.

80.9% Most of the time.

1.7% When budget and schedule permit.

MQ-17. On whether electrical incidents are reported:

36.6% Always reported.

60.7% Reported only if an injury results.

2.7% Seldom reported even if an injury results.

0% Rarely reported under any circumstances.

MQ-18. On whether their personnel who are exposed to potential electrical hazards have attended an HSE-3 electrical safety course:

17.0% All have attended.

34.9% Most.

22.6% Some.

25.5% None.

MQ-19. On whether it is necessary for their personnel to work alone on exposed, energized electrical equipment:

27% yes.

73% no.

MQ-20. On whether managers have a set of guidelines for working alone on electrical equipment:

80% yes.

20% no.

V. GENERAL FINDINGS

The findings of the Committee are based on the results of the two questionnaires, 70 interviews with employees throughout the Laboratory, and the experience and opinions of the individual Committee members. The questionnaires were generated and analyzed in a relatively short time and the Committee is fully aware of their shortcomings with respect to a rigorous and completely defensible statistical analysis. However, our intent was to obtain, as quickly as possible, a general flavor of Laboratory opinion and advice, and to that degree the Committee believes it succeeded. The raw data in appendixes provide fertile ground for future analysis.

Before proceeding with findings and recommendations, it should be stated that the problem areas described are, of course, not equally applicable to all organizations. The

Committee found that some groups appear to be doing a very creditable job of implementing proper electrical safety policy. Predictably, these organizations are commonly those whose principal occupation is the construction and operation of potentially lethal electrical equipment. This probably stems from both an availability of skilled personnel, and, more importantly, a high degree of electrical safety consciousness. Instilling this consciousness throughout the Laboratory should be a major goal. The results demonstrate that the organization's attitude strongly affects the degree to which individuals follow safety requirements.

A. Laboratory Policy

The Committee believes that the Laboratory's general safety policy and specifically the electrical administrative requirements are good. The Health, Safety, & Environmental Division's Manual provides the necessary framework for an excellent safety program. Many of the recommendations in Section VI are restatements of requirements already in the HSE Manual.

One of the basic tenets of Laboratory safety policy is that no job will be performed if it cannot be done safely. There appears to be a significant number of electrical workers who are either unaware of this policy or believe that it cannot always be followed in practice.

The Committee found that both the letter and spirit of electrical safety policy is applied in a very nonuniform manner around the Laboratory. While rationalizations may be given such as personnel limitations, schedule constraints, or "impracticality," the underlying problem seems one of attitude.

- For whatever reasons, some individuals at the hands-on working level seem to accept electrical hazards as "part of the job." Risks are taken that would be completely unacceptable in other areas of activity. Somehow the hazards of chemical toxins, radiation, or explosives seem more "real."

- Allied to the above, the Committee found that some managers do not rigorously apply electrical safety requirements in their operations. Attitudes of passivity or tolerance towards electrical risks must be changed, and Division and Group Leaders are the key people here. Special attention is needed in groups whose primary discipline is "non-electrical" but who nonetheless use a large amount of potentially hazardous electrical equipment.

The Committee also notes that recommendations similar to our own were made by the Ad Hoc Committee on Electrical Safety in a report dated September 1974. As far as the Committee has been able to determine, a follow-up to that report was not affected Laboratory wide.

Relevant Data:

Q-10. First ranked cause of accidents is rushed or distracted personnel.

Q-15. 36% of electrical workers believe that electrical safety is left to individual judgment or is not considered of first importance in their organization.

Q-16. 53.2% of electrical workers are not familiar with HSE's Manual (AR7-1) requirements.

Q-19. 36.2% of electrical workers who expressed an opinion believe that people in their organization work alone on exposed, energized electrical equipment 50% or more of the time.

Q-20. Given that people worked alone on energized equipment, the ranked reasons were:

1. Did not think it was necessary to have a second person.
2. Could not get a second person.
3. Group/Section puts other concerns first.

MQ-8. 50% of the managers have not formally evaluated their work places for electrical hazards.

MQ-9. Only 35% of managers review their SOPs/SWPs at least annually.

MQ-19. 27% of the managers find it necessary for their personnel to work alone on exposed, energized equipment.

B. Communications Problems

It is concluded that both vertical and horizontal communication channels need improvement. Employees must feel completely free to report, either formally or informally, all incidents, perceived unsafe conditions, or criticisms of policy without fear of consequences. Managers must see that general Laboratory policies and requirements are clearly explained and followed in their organizations. Where general policies or requirements are inadequate or deemed impractical, alternate written procedures giving *equal* protection should be developed. An area that needs much improvement is horizontal communications within organizations to disseminate information on operating manuals, as-built drawings and schematics, changes to equipment, power and equipment status, etc. (Historically, a prominent common factor in Laboratory electrical accident and incident reports is lack of accurate equipment documentation.)

It was found that many electrical workers were unaware of the new electrical Administrative Requirement AR7-1, and there was considerable confusion between this requirement and the old AR14-1. As the new HSE Manuals have not been issued, many thought that the new requirement was numbered incorrectly. In many organizations, it appears that new requirements simply get filed in a Group Office Manual and the information is not always transmitted to people doing the work.

Relevant Data:

Q-9. 15.2% of electrical workers believe that incidents are reported 1/3 of the time or less.

Q-16. 53.2% of electrical workers are not familiar with the requirements of AR7-1.

Q-23. Causes of power systems problems were ranked:
1. 25% — Improper wiring of plugs and outlets, and 25% — lack of standards for power connections.
2. 19% — Inadequate drawings and schematics.
3. 15% — Poor inspection, and
   15% — Inadequate maintenance.

MQ-11, 12, & 15. 81% of managers organizations modify equipment.

24% always document modifications.

56% have a communications system for informing personnel of critical electrical design information.

MQ-13 & 14.   40% of managers do not routinely apply AR7-1 in their work

51% have not distributed AR7-1 or discussed it with their people.

MQ-17.   60.7% of managers believe that incidents are reported only if an injury results.

C. Conflicting Pressures

A significant portion of electrical workers and managers feel that safety considerations are compromised by the pressures of schedule, budget, and personnel limitations. While it is proper to make first-line managers directly responsible for providing a safe working environment, higher management must also indicate that it recognized the costs of a comprehensive safety program. Whether expressed or implied, two themes were common in the questionnaires and interviews:

- "I do not have the resources to adhere to safety policy at all times," and
- "As a manager, I an primarily evaluated on whether I meet schedules and budgets."

Relevant Data:

MQ-4.   33% of managers believe that the Laboratory's safety program is inadequate or does not address their specific needs.

MQ-10.   Managers ranked reasons that may require compromise with safety policy.

1. Other. (Policy is inflexible, inconsistent, or irrelevant (4), employee lack of interest in safety (4), identifying problem areas (3), remote field areas (3), our safety is not compromised (12).
2. Schedule constraints.
3. Insufficient budget.
4. Insufficient emphasis by upper management.

D. Lack of Electrical Standards

Comments on the lack of standards for power connections and color coding were almost universial with respondents who worked in this area. There is much confusion and considerable finger pointing between Laboratory users, the Zia Company, and the Engineering Division. In some areas, there is confusion or disagreement over interfaces between utility and user power. A number of serious incidents have occurred as a direct consequence of the situation. While the National Electrical Code provides general guidance on plugs, outlets, and color coding, it is left to the using organization (the Laboratory) to adopt specific options.

The Laboratory also does not have procurement standards for the purchase of off-the-shelf or custom fabricated electrical equipment.

Relevant Data:

MQ-3. Managers ranked the most significant potential electrical hazards:

1. Power circuits over 100V.
2. Power supplies.
3. Capacitors.
4. Instruments and Control Systems.
5. Pulsed Power.
6. Electrical conductors and connectors.

MQ-5. Managers ranked the three most prevalent hazards:

1. Machinery and other.
2. Electrical.
3. Lasers.

E. Training Programs

There is need to improve both the content and number of electrical safety training courses. The most common negative comments were that the HSE-3 (Industrial Safety) course was boring or that it was not relevant to the respondent's work. However, these comments in themselves indicate a misunderstanding of responsibility. The using groups are responsible for developing (with HSE assistance) electrical safety training programs that are relevant to their work. It should be pointed out that much of the negative comment was based on the old HSE-3 safety course. HSE has modified the old course and has been presenting it for several months. The new course addresses many of the criticisms, and HSE-3 is continuing to improve it with input from the using groups.

Relevant Data:

- **Q-6.** 60% of the 40-600V workers and 45% of the over 600V workers have not had an HSE electrical safety course.
- **MQ-6.** 25.7% of managers responding do not have an electrical safety policy specifically addressing their hazards.
- **MQ-18.** 51.9% of managers say all or most of their personnel who are exposed to electrical hazards have attended an HSE electrical safety course.

F. Reporting of Incidents and Accidents

There is evidence that most minor electrical incidents (and some not so minor) do not get reported even formally. This finding is closely related to the one on communications. Even minor shock incidents are useful as a discussion subject within organizations, and personnel should feel no inhibitions in reporting them. The SM/TEC questionnaires indicate that at least 173 people received electrical shocks or burns in the past year at the Laboratory. The actual number is probably far larger.

Relevant Data:

- **Q-7 and 8.** 21.4% of the electrical workers reported one or more electric shocks in the past year; 1.3% reported burns.

G. Comparison of Management and Nonmanagement Responses

In a few areas of electrical safety, the committee compared responses from the nonmanagement SM/TEC questionnaires and the managers. Direct correlation in some instances is difficult as the questionnaires were not originally designed for parallel comparison. Comparison plots were made, however. In summary, the plots generally indicate:

- A larger fraction of managers say they routinely apply AR7-1 in their work than the percentage of workers who say they are required to be knowledgeable of AR7-1.
- A larger fraction of managers say they have a communication system for informing personnel of critical electrical design information than workers who say they always receive such information.
- Both management and nonmanagement responses indicate that more personnel need to take the HSE-3 electrical safety course.
- A much higher fraction of managers say they do not find it necessary for their personnel to work alone on exposed, energized, electrical equipment than workers who say they never do such work.
- Both management and nonmanagement responses indicate that only a fraction of electrical indents are always or almost always reported.

VI. RECOMMENDATIONS

A. Policy

1. Managers, especially Group and Division Leaders, must be convinced that Laboratory policy is to be taken literally, that is, every job must be done safely. At the working level, the Group Leader's attitude sets the time for the entire organization. Two approaches to this recommendation are suggested for consideration:

   a. The existing safety appraisal system utilized by HSE Division might be augmented to provide an organizational safety rating factor that could be utilized in evaluating responsible managers.
   b. A safety program evaluation system applicable to groups and divisions could be devised that would utilize HSE appraisals, accidents histories, and other factors to reward responsible managers.

The Committee recommends that HSE Division (with others) be tasked to establish a uniform method of safety program evaluation applicable to groups and divisions. The system should provide organizational safety ratings which will be an important factor in evaluating responsible managers.

2. Division Leaders must provide the resources for their groups to carry out safety policy properly. In some instances this may require additional personnel or schedule adjustments. Safe operations are an essential part of doing business, and Division Leaders should make it clear that the pressures of budgets or schedules do not justify compromising safe practices. This philosophy appears better established in the Laboratory's Special Nuclear Materials, radioactive materials, and explosives operations than in electrical work.

3. Individual group members in turn must be convinced that proper safety procedures are realistic and will be enforced. Mutually supportive training efforts between the group managers and HSE Division are essential. The policy of either safe operations or no operations should be emphasized at every safety meeting.

4. The Committee recommends establishment of a permanent Electrical Safety Committee under ther Director's Safety Council. A similar recommendation was made by the Ad Hoc Committee on Electrical Safety in 1974, and it is quoted directly:

> "The committee believes the time has come for the Director to appoint an Electrical Safety Council. This council would be patterned after other Director's safety committees, the committee on nuclear criticality, for example. "It is suggested that this Electrical Safety Council serve in the following ways:
>
> 1. The council would assist in the formulation of laboratory-wide policy on safe practices in electrical technology.
> 2. The council periodically would make spot investigations to see if the policies on electrical safety were functioning as intended and serving the Director's will.

3. The council would serve singly or as a group when requested by the various technical divisions and by the Safety Group, H-3, to provide expert advice on matters of electrical safety.
4. The council members would be made up of people with a primary interest in electrical technology. They would not be professional safety personnel, and, to prevent their having an excessive safety workload, the committee would not be asked to participate on routine safety reviews."

This Committee agrees with the essence of the recommendations and adds that the Committee would assist in the development of electric power standards for Laboratory facilities and procurement standards for electrical equipment.

B. Communications and Controls

1. Groups should be required to have safety plans (SOPs or SWPs) for potentially hazardous electrical work. These should be both general (e.g., for working on exposed energized electrical equipment) and specific for particular pieces of equipment or operations. There is confusion in some organizations between SOPs and Operating Instructions. This should be resolved. Operating Instructions do not substitute for SOPs.

2. Groups should be required to maintain complete and up-to-date documentation on all their equipment in a central file known to all operating or maintenance personnel in the group. Where applicable, copies should be in a pocket attached to individual apparatus. A maintenance log should also be in this pocket.

3. It should be made clear that the Manual or sections of the manual are available to anyone with a legitimate need. (A few technicians commented that they could not obtain a Safety Manual for their workplace.) HSE Division should continue to maintain expanded distribution lists for critical information on electrical safety. These lists currently use the Master Management and Safety Chairperson listings for all groups.

4. Groups that perform electrical work or that use a significant amount of electrical equipment should be required to have a formal system for disseminating critical electrical information within the group. For example, frequent (at least monthly) meetings of staff and technicians to discuss new equipment, changes that affect equipment safety, power status, minor incidents, etc. Groups should also ensure that safety information receives wide distribution. For example, AR7-1, which was sent to all HSE Manual holders, should have been distributed by each group to all personnel responsible for maintaining and operating electrical systems.

5. Groups should be required to perform electrical safety inspections on new equipment or on equipment that has been electrically modified before placing it in service. These inspections should be recorded in the maintenance log attached to the equipment.

C. Electrical Standards

1. The facilities Engineering Division has begun to develop and issue a set of electrical standards for power wiring including color coding, plugs, and outlets for the Laboratory. The Engineering Division (with HSE and others) should be tasked to expedite formalization and issuance of any pending electrical standards and assure that standards are regularly reviewed and revised where required. These standards should be mandatory for all new construction and for retrofitting older facilities as time and budgets permit. The standards should be based on DOE Orders and the National Electrical Code. The essential points are that they be uniform for the Laboratory, rigorously implemented, and maintained up-to-date.

2. The Committee agrees with the position of ENG Division and the Zia Company that using groups should be prohibited from modifying utility wiring. The utility/nonutility interfaces should be clearly defined and a directive to this effect issued. Where equipment is not compatible with facility wiring, the equipment should be modified or an adapter made. The ENG Division and the Zia Company must be geared to respond promptly to work requests if such a prohibition is to be effective.

3. The Laboratory or the Zia Company should stock the necessary hardware to accommodate new construction and a continuous program of retrofitting to the Laboratory's electrical power standards. They must also stock hardware to enable fabrication of the above-mentioned adapters.

4. All facility power outlets in the Laboratory should be inspected for voltage, phasing, and grounding, and clearly marked with permanent labels.

5. The Laboratory should develop a set of basic procurement standards for electrical apparatus. This will be a sizeable task and might best be done by a consultant. The Committee has a set of procurement standards developed by the IBM Corporation that might serve as a guide. Such standards address, among other things, quality of components, safety devices, and wiring conventions.

D. Training

1. A common complaint was that existing electrical safety training programs were rather general and not always directed to the specific needs of a particular group or section. HSE-3 has been and is continuing to improve the electrical safety course. This course is designed for general electrical hazard (nonjob-specific) and Laboratory requirement training. It is intended to increase individual awareness of potential electrical hazards. Groups performing electrical work should (with HSE assistance) develop their own job specific electrical safety courses that are geared to the potential hazards their personnel are exposed to.

2. A number of comments from technicians were to the effect that they would like to know more about the theory and operation of the equipment they are involved with. Staff and senior technicians should hold regular training sessions for exchanging such information.

8
Instituting an Effective Electrical Safety Management Program in Research and Development Facilities

The following information was extracted from a report, "Electrical Safety Criteria for Research and Development Activities," by the U.S. Department of Energy, published in 1979.

The ability of the human body to withstand safely the internal passage of electrical current is so small that most electrical systems can be hazardous to health and life. The hazards of conventional residential, commercial, and industrial electrical systems and equipment are generally well recognized; and appropriate codes, standards, and safety regulations have been established for application in the design, construction, and operation of this equipment. However, special hazards exist in research and development (R&D) activities involving the use of unique electrical and electronic systems, specialized equipment, and unusual use of conventional equipment. Existing codes, standards, and safety regulations often do not provide adequate coverage for these applications.

The special electrical safety hazards associated with R&D activities at the Department of Energy laboratories have been, and continue to be, of serious concern. During the 16-year period from 1962 through 1977, there were over *200 reported electrical accidents* directly associated with R&D equipment and activities at these laboratories (formerly laboratories of the prior Atomic

Energy Commission (AEC) and the Energy Research and Development Administration (ERDA)). About *forty percent were serious* accidents involving moderate-to-severe electric shock and/or body burns. There were *three* laboratory employee electrocutions. The remainder of the accidents involved minor shock and/or minor body burns.

In recognition of the special electrical hazards, the AEC Division of Operational Safety established an ad hoc electrical safety committee in 1966, with technical representation from the AEC Division of Construction and seven major AEC laboratories. That committee developed "Electrical Safety Guides for Research," published in 1967 as AEC Safety and Fire Protection Bulletin No. 13. In 1975, the ERDA Division of Operational Safety reconstituted the committee to update and further develop electrical safety criteria for R&D, with technical representation from the ERDA Division of Facilities and Construction Management and eleven of the ERDA (now Department of Energy) laboratories. This committee developed "Electrical Safety Criteria for Research and Development Activities," that provide:

1. General criteria to be followed for laboratory electrical safety administration;
2. General safety criteria to be followed in the design, construction, and operation of electrical R&D systems and equipment;
3. Safety criteria to be followed in working on electrical equipment; and
4. Additional safety criteria to be followed in the design, construction, and operation of those categories of electrical equipment commonly used in R&D activities.

Because the report covers the information needed for this chapter, it is presented in its detailed style below.

SCOPE AND APPLICATION (Part I)

Scope

Department of energy (DOE) laboratories engage in a variety of R&D activities which often involve new or unusual electrical equipment not sufficiently covered by established electrical codes or standards. Therefore, special efforts must be made to assure adequate electrical safety—beginning with the design phases and continuing through development, construction, and operation of R&D equipment and facilities.

1. Due to the differences in R&D program requirements at DOE laboratories and the unpredictability of R&D developments, it is not practical to establish a specific set of safety program requirements to be applied uniformly. There are, however, general safety criteria applicable to all DOE laboratories, and these are contained in Parts II through VI.

2. Part VII includes safety exhibits for some specific categories of electrical equipment which are common to most DOE laboratories. Each exhibit contains applicable design, construction, and operating criteria.

3. Part VIII describes other types of hazards and contains general design, construction, and operating criteria for dealing with nonelectrical problems associated with the R&D equipment covered in the Part VII exhibits.

4. *Appendix A* contains definitions of special terms used in these Criteria. *Appendix B* contains a bibliography of useful references. *Appendix C* lists the membership of the Electrical Safety Criteria Committee.

Application

1. These criteria are for interim application by DOE laboratories. It is planned to finalize the criteria following a six- to nine-month period of use, incorporating appropriate suggested revisions or additions.

2. The DOE Division of Operational and Environmental Safety plans to adopt the final criteria as a recommended standard to be followed in electrical safety administration and operations at the DOE laboratories.

GENERAL CRITERIA FOR ELECTRICAL SAFETY ADMINISTRATION (Part II)

This part covers the general criteria for electrical safety administration and organization which are applicable to all DOE laboratories in scope and intent. These criteria are for use by each laboratory as a basis for preparation and implementation of a strong program for electrical safety.

Administrative Policy

1. The responsibility for safety rests ultimately with the Head of each Laboratory. Consequently, and effective safety program begins with his statement of safety policy. This statement shall include the reasons for and objectives of the safety program, with due emphasis on electrical safety.

2. The lines of safety responsibility and authority which extend from the Head of the Laboratory to every person employed by the Laboratory, or using the Laboratory facilities, shall be clearly established. The formal safety organization and special electrical safety committees shall be included.

3. Provisions for handling exceptions or variances from the established rules and procedures shall be clearly defined.

Safety Organization and General Program Requirements

1. Each Laboratory shall establish a safety organization at a relatively high level within the Laboratory's structure with adequate funding to develop and implement the safety programs, giving due emphasis to electrical safety. Persons appointed to this organization shall be of sufficient professional stature and delegated the necessary authority to assure that safety is not compromised for expediency of experiments or operations. The structure of this organization will depend upon the safety program requirements of the particular laboratory.

2. One person shall be assigned the first-line responsibility and the necessary authority for the safety of each R&D operation or activity in accordance with all applicable DOE and Laboratory safety policies and procedures.

3. Each new research program and major revision of an existing program, or other revision that could affect safety, shall be reviewed by the Laboratory, with particular emphasis on electrical safety features.

4. Periodic electrical safety reviews of existing research programs and facilities shall be conducted to assure that cumulative modifications and additions have not introduced new unprotected hazards or negated the protection previously provided. Such reviews shall include the actions and

procedures used by all personnel involved with the program.

5. An internal audit procedure shall be established for periodic self-evaluation of the effectiveness of the Laboratory's electrical safety program. The evaluation shall include written reports of accidents, incidents, local safety problems, and necessary corrective actions taken or planned.

6. Basic electrical safety requirements shall be established and implemented to minimize the risks associated with all R&D activities.

7. An emergency plan shall be in effect at all times to cover adequately potential electrical accidents or incidents involving personnel or equipment. Laboratory staff, experimenters, and other personnel shall be indoctrinated as to their responsibilities, authorities, and required actions in the event of emergency or equipment malfunction.

8. A sufficient number of personnel who frequent hazardous areas shall receive first-aid training, including cardiopulmonary resuscitation (CPR), so that assistance is readily available in the event of emergencies.

9. All employees should be given a training course in electrical safety which is commensurate with their work assignments. Those actively engaged in research or related support activities should receive a refresher course at regular intervals. Electrical safety problems and hazards should be identified and discussed at periodic safety meetings.

10. Each Laboratory should prepare an Electrical Safety Manual for use when training employees in the performance of their assigned duties, and for reference. The manual should contain the Laboratory's electrical design, construction, and operating criteria, and maintenance safety rules and procedures; emergency plans and procedures; and criteria and procedures for work on electrical equipment commensurate with the criteria contained in Part V.

Codes, Standards, and Criteria

Laboratory safety administration shall assure, to the maximum extent practicable, that electrical and electronic systems and equipment design, construction, and operation conform to these Electrical Safety Criteria and to current editions of other applicable codes, standards, and criteria including:

1. The National Electrical Code (ANSI C1 or NFPA No. 70 of the National Fire Codes, National Fire Protection Association);

2. National Electrical Safety Code (ANSI C2);

3. Occupational Safety and Health Administration (OSHA) Standards and Regulations, Parts 1910 amd 1926;

4. National Fire Codes of the National Fire Protection Association;

5. Other applicable standards, criteria, and guides contained in Appendix B (Bibliography) of these Electrical Safety Criteria; and

6. Other, more stringent, criteria as Laboratory Management deems necessary.

GENERAL SAFETY CRITERIA FOR DESIGN AND CONSTRUCTION (Part III)

This part contains general safety criteria to be followed by DOE laboratories in the design and construction of electrical R&D equipment and facilities. Additional criteria for specific types of electrical R&D equipment are contained in Part VII.

A. Provide sufficient access and work space around electrical equipment for personnel safety during operation and maintenance.

B. Provide adequate illumination and automatic emergency lighting in all R&D areas where electrical hazards may be encountered.

C. Provide adequate ventilation in areas where electrical faults may produce noxious fumes and where high voltages or radiation may produce ozone.

D. Provide adequate floor drainage around electrical equipment where water accumulation could occur, to reduce electric shock hazards.

E. Provide physical barriers to prevent personnel or hand-held electric-conducting material from contacting energized parts. Enclose equipment that operated at greater than 50 volts.

F. Provide protective covers or barriers for high-voltage terminals, and for low-voltage terminals at which high currents are available.

G. Provide first-line and backup protection where serious electrical hazards exist, to prevent access to energized circuits and parts.

H. Identify hazardous areas and the nature of each hazard by such means as warning signs, flashing lights, or audible alarms.

I. Design protective devices, equipment, and/or systems to be fail-safe wherever practicable.

J. Ground metal cabinets, enclosures, and structural components and equipment with easily recognizable, permanent, external grounding conductors adequately sized for the maximum available fault currents, and having sufficiently low impedances to assure that hazardous voltages are not developed when the maximum currents are flowing in the grounding conductors.

K. Provide safety grounding hooks for hazardous electrical R&D equipment with:

1. A bare, extra-flexible copper ground conductor of #2-AWG size (minimum);
2. Connectors securely fastened to the flexible ground conductor; and
3. The bare conductor clearly visible through its insulating sheath.

Ground conductors shall have a clearly visible metal-to-metal bolted connection to equipment ground. Grounding hooks shall be of sufficient number to satisfy the equipment grounding requirements, and shall be located in a clearly visible and accessible location (see definition in Appendix A).

L. For equipment that may retain residual charge (e.g., capacitors or capacitor banks) after the primary power source has been deenergized, provide a discharge point (or points), in addition to a direct grounding point (or points). Clearly label all discharge and direct grounding points. Provide discharge points with impedances capable of limiting the discharge current to 50 amperes or less.

M. Provide means of communication (e.g., telephone or intercom) near hazardous electrical equipment for use during emergencies. Label communication equipment to identify clearly its location so that proper instructions can be given to personnel responding to an emergency.

N. Distinctively color-code and/or label any component which in its common use is nonhazardous but in its actual use may be hazardous (e.g., a metallic cooling-water pipe also used as an electrical conductor, carrying high voltage or high current).

O. Install emergency power-shutdown switches at strategic locations to facilitate prompt deenergizing of electrical power supply to hazardous R&D equipment. The emergency shutdown switches shall be clearly labeled for unmistakable identification and shall be readily accessible and physically unobstructed.

P. Provide a main power disconnect (disconnect switch or circuit breaker) for hazardous electrical equipment at the power distribution panel capable of being padlocked in the disconnect (open) position.

Q. Clearly label all disconnects and cirsuit breakers as to the loads they control, especially if they service high-hazard equipment. Where disconnect switches are used, labeling shall include identification as to their load-break capabilities (e.g., "This disconnect switch has _____ (amperes/KVA) of load-break capability," or "This disconnect switch *does not have* load-break capability and *must not* be operated under load conditions"). Similarly, label all hazardous equipment with the identification and location of the servicing power-disconnect switch or circuit breaker.

R. Provide warning or pilot lights to indicate clearly that equipment is energized.

GENERAL OPERATING SAFETY CRITERIA (Part IV)

This part contains general safety criteria to be followed by DOE laboratories in the operation of electrical R&D equipment and facilities. Additional operating safety criteria for specific types of electrical equipment are contained in Part IV.

A. Maintain up-to-date drawings and instructions for operation, maintenance, and testing of electrical equipment and assure their availability to personnel working on hazardous equipment.

B. Before beginning work on hazardous electrical equipment adequate lock-and-tag procedures shall be employed.

C. Personnel working on hazardous electrical equipment shall use protective equipment appropriate for the exposure (e.g., safety glasses, insulating gloves, and insulating mats).

D. Whenever anyone is working on hazardous electrical equipment, a second person capable of helping in an emergency shall be present. See Part V for details.

E. Manual protective ground devices, such as grounding hooks, shall be attached to hazardous electrical equipment after the equipment has been deeenergized and before personnel are permitted to work. Grounding devices shall be left connected in place until the work is finished. Where discharge points are provided in addition to direct grounding points, protective ground connections shall first be made to the discharge points to assure that residual stored energy is safely dissipated before making protective ground connection to direct grounding points.

F. Anyone doing hazardous work who appears to be fatigued, ill, under the influence of alcohol or drugs, or emotionally disturbed, shall be replaced by a competent backup person; otherwise the hazardous work shall be terminated.

G. Safety interlocks and interlock systems shall be tested at least annually to assure operability; the use of written check lists is recommended. Electrical interlocks and other safety devices shall not be bypassed except when absolutely necessary. Procedures shall be established and implemented for obtaining approval, tagging the interlock, and logging its location and the times it is bypassed and restored.

H. When hazardous electrical equipment is being serviced, temporary physical barriers and warning signs shall be used for personnel protection.

I. When control circuits remain energized after the primary power source to electrical equipment is disconnected, the equipment shall be clearly labeled as to the condition.

J. When electrical equipment is being serviced, supplemental illumination shall be provided as required, and the immediate area around the equipment shall be uncluttered and unobstructed.

K. Persons working on hazardous electrical equipment should never wear metallic items (e.g., key chains, wristbands, watchbands, or rings). In addition, personnel should also be made aware that metal flashlights and other working tools may also introduce safety hazards.

SAFETY CRITERIA FOR WORKING ON ELECTRICAL EQUIPMENT (Part V)

This Part contains essential criteria to be followed by DOE laboratories in developing procedures for work on electrical equipment used in R&D. Special emphasis is placed on adequate planning, hazards recognition, and administrative requirements. Safety requirements for work on conventional industrial/commercial electric systems and equipment are generally satisfied by existing industrial safety practices and procedures and are not specifically addressed here.

A. Minimizing Risk

Minimizing the risk of accidents is the fundamental criterion to be satisfied whenever there is work on R&D electrical equipment. This means working on equipment in its positively deenergized state and this approach should be followed by DOE laboratories to the maximum extent practicable. Work on energized equipment, with interlocks defeated or protective barriers removed, should be permitted only as a last resort after all reasonable attempts have been made to work with the equipment deenergized.

B. Planning the Work

1. Routine work shall be planned carefully and scheduled sufficiently well in advance to minimize adverse impact on the R&D project or program.

2. Where emergency work is required, electrical safety shall not be compromised in favor of maintaining continuity in R&D activities. Equipment outages shall be scheduled in lieu of working on energized equipment, wherever possible.

3. When planning routine or emergency work, identify all associated and potential electrical hazards, including built-in (inherent) equipment hazards. Advise personnel assigned to the work of these hazards the procedures to be followed, what protective measures to be taken during work performance.

C. Classes of Electrical Hazards

It is of paramount importance to identify the hazards associated with each type of electrical equipment used in R&D activities when developing and applying procedures administrative controls for work on the equipment. Ideally, "standard" electrical hazard classifications might be developed for all types of equipment. If supply voltage were the only hazard parameter, this would be possible. However, because of the many possible types of electrical hazards in complex R&D equipment, it was not feasible to establish "standard" hazard classifications for such types of equipment.

For purposes of these criteria, from practical and psychological considerations, electrical hazards may be divided into two general classes, identified herein as *Class A* and *Class B*. *Class A* is the *lesser hazard* situation where, based on judgment on experience, the electrical hazards are such that they can usually be safely controlled with a minimum of personnel and administrative requirements. *Class B* hazards are identified as those where any of the Class A limits are exceeded.

NOTE

It is recognized that there cannot be an exact dividing line. Judgment must be used in applying these classifications.

"Class A" Hazard

A "Class A" electrical hazard exists when *all* of the following four conditions are satisfied:

1. The nominal primary (equipment source) ac voltage does not exceed 130 volts (rms value).

2. The available primary ac load current is limited to 30 amperes (by supply circuit breakers or fuses having load ratings not exceeding this value).

3. The stored energy available in a capacitor or inductor does not exceed 10 joules.

4. The dc voltages (input or output) or internal secondary ac voltages present in instrumentation and test equipment *either* do not exceed 30 volts between terminals or to-ground, *or* have power limits of 150 volt-amperes.

While *Class A* voltages, currents, and energies may be considered small, a *Class A electrical hazard* must be regarded as potentially lethal. This Class is particularly dangerous because of its everyday familiarity and an assumed ability to cope with its common occurrence in exposures guarded to a lesser extent than in higher hazard situations.

"Class B" Hazard

A "Class B" electrical hazard exists where *any one* (or more) of the following four conditions is present:

1. The nominal primary (equipment source) ac voltage exceeds 130 volts (rms value).

2. The available primary ac load current exceeds 30 amperes (with supply circuit breakers or fuses having load ratings in excess of 30 amperes).

3. The stored energy available in a capacitor or inductor exceeds 10 joules (at about 100 volts or more).

4. The dc voltages (input or output) or internal secondary ac voltages present in instrumentation and test equipment exceed 30 volts between terminals or to-ground *and* have power limits greater than 150 volt-amperes.

D. Modes for Working on Electrical Equipment

The following three modes of operation will serve as an aid in establishing requirements for working on electrical equipment.

Mode 1—All operations are to be conducted with the equipment in a "positively deenergized" state. All external sources of electrical energy are disconnected by some positive action (e.g., locked-out switch or circuit breaker) and all internal energy sources are made safe (discharged and grounded).

"Mode 1 is a minimum hazard situation which requires little supervision of the worker(s) after the initial protective steps are taken.

Mode 2—All manipulative operations (such as making internal electrical connections or equipent alterations) are to be conducted with the equipment in a "positively deenergized" state. Equipment functions may then be observed *remotely* after the equipment has been reenergized, with normal protective barriers removed.

"Mode 2" is a moderate-to-severe hazard situation, depending on the operating voltages and energy capabilities of the equipment and its power source and the complexities of the work to be performed.

Mode 3—Manipulative, monitoring, and observational operations are to be conducted with the equipment energized and with normal protective barriers removed.

"Mode 3" is the most severe hazard situation and should be permitted only when fully justified and with the operations conducted under close supervision and control.

General Safety Considerations

1. Develop and utilize written procedures for authorizing work on hazardous equipment.
2. Assign an adequate number of "knowledgeable persons" for work on each piece of hazardous equipment.
3. Assign a "safety watch" person whenever the complexities of the work and the associated electrical hazards dictate.
4. Perform all work feasible with the equipment in a "positively deenergized" state, employing lock-and-tag procedures.

F. Minimum Personnel and Administrative Requirements

Personnel and administrative requirements to aid in minimizing the risks in work on electrical equipment are dependent on the Class of Electrical Hazard and Working Mode. Recommended "Minimum Requirements" to be followed for work on electrical equipment are contained in the Table 1.

In applying these requirements, judgment must be used. As stated in C. above, these cannot be an exact dividing line between hazard classification (*Class A* vs. *Class B*). Also, there may be situations where some deviation from the requirements can be justified. For example, it is recognized that the internal voltages present in many digital types of equipment and integrated circuit (IC) equipment/systems are sufficiently low (e.g., below 30 volts) as to present minimal electric shock/burn hazards. Safety experience in working on such equipment has been favorable in industry, universities, and DOE laboratories; and consideration should be given to permit personnel to work alone on such equipment in the energized state. However, before deviations from these requirements are permitted, hazard evaluations (and risk analyses) shall be made. Deviations granted should be on a case-by-case basis. A blanket deviation for a particular category of equipment should only be granted after all of the potential electrical hazards have been evaluated, and this type of deviation has been fully justified.

ELECTRICAL HAZARDS AND EMERGENCY PROCEDURES
(Part VI)

A. Electrical Hazards

1. The ability of the human body to withstand internal passage of electrical current (through the heart region) is so small that most electrical systems can be hazardous to health and life (see Table 2).

2. Even minor electrical shock can create serious injury due to a bump, trip, or fall resulting from reflex body action.

3. Exposure to as little as 0.8 joule per cm^2 can cause minor thermal burns on the skin. As the greatest resistance of the body is the skin, that is where burns are most likely to occur on electrical contact. The thresholds of electrical burns caused by current passing through tissue are difficult to establish, since

Table 8.1 Minimum Personnel and Administrative Requirements for Work on Electrical Equipment

Electrical Hazard Classes	Working Modes		
	Mode 1 (Minimum Hazard Situation) All operations performed with equipment in a "positively deenergized"[a] state.	Mode 2 (Moderate-to-severe Hazard Situation) Manipulative operations performed with equipment in a "positively deenergized" state. Equipment functions are then remotely observed and monitored with equipement energized and normal protective barriers removed.	Mode 3 (Severe Hazard Situation) Manipulative, observational, and monitoring operations performed with equipment energized and with normal protective barriers removed.
Class A (All of the following four conditions are satisfied) 1. The nominal primary (equipment source) ac voltage does not exceed 130 volts (rms value). 2. The available primary ac load current is limited to 30 amperes (by supply circuit breakers or fuses having load ratings not exceeding this value). 3. The stored energy available in a capacitor or inductor does not exceed 10 joules. 4. The dc voltages (input or output) or internal secondary ac voltages present in instrumentation and test equipment *either* do not exceed 30 volts between terminals or to-ground, *or* have power limits of 150 volt-amperes.	After equipment has been "positively deenergized," personnel may work alone with a minimum of supervision. *A-1*	Personnel may work alone on assigned tasks with general supervision. (See Note 1) *A-2*	Personnel may work on assigned tasks with general supervision, however a companion[a] shall be required. (See Note 1) *A-3*
Class B (*Any one*, or more, of the following conditions are present) 1. The nominal primary (equipment source) ac voltage exceeds 130 volts (rms value). 2. The available primary ac load current exceeds 30 amperes (with supply circuit breakers or fuses having load ratings in excess of 30 amperes). 3. The stored energy available in a capacitor or inductor exceeds 10 joules (at about 100 volts or more). 4. The dc voltages (input or output) or internal secondary ac voltages present in instrumentation and test equipment exceed 30 volts between terminals or to-ground *and* have power limits greater than 150 volt-amperes.	Two knowledgeable persons[a] required until the "positively deenergized" state is clearly established. Then, personnel may work alone on assigned tasks with general supervision. *B-1*	Two knowledgeable persons at all times, plus verbal approval by first level supervisor. (See Notes 2 and 3) *B-2*	Two knowledgeable persons plus a safety watch.[a] (See Note 3) Written approval for the specific operations to be performed shall be required from the first and second levels of technical supervision and from appropriate safety officer. *B-3*

[a]See Appendix A for definition of terms.

Notes: 1. In Class A, Modes 2 and 3 situations, work in confined spaces or under conditions involving massive grounds* shall require two knowledgeable persons.
2. In Class B, Mode 2 situations, when a high degree of hazard exists in the equipment itself, or may result from repetitive sequences of manipulative-observational operations that increase the risk of human error, written approval and a safety watch shall be required.
3. In Class B, Modes 2 and 3 situations, the two knowledgeable persons shall discuss and agree on each overt act, and only one person at a time will perform an act.

Table 8.2[a] Current Range and Effect on 150-lb Man

Current (60 Hz)	Physiological phenomena	Feeling or lethal incidence
•1 mA	None	Imperceptible
1 mA	Perception threshold	
1-3 mA		Mild sensation
3-10 mA		Painful sensation
10 mA	Paralysis threshold of arms	Cannot release hand grip. If no grip, victim may be thrown clear. (May progress to higher current and be fatal)
30 mA	Respiratory paralysis	Breathing stops. (frequently fatal if not treated promptly)
75 mA	Fibrillation threshold 0.5%	Heart action is discoordinated (probably fatal)
250 mA	Fibrillation threshold 99.5% (5-s exposure)	
4 A	Heart paralysis threshold	Heart stops during current passage, restarts normally on current interruption (usually not fatal from heart disfunction)
□5 A	Tissue-burning	Not fatal unless vital organs are burned

[a]R. H. Lee, Electrical Safety in Industrial Plants, IEEE Transaction on Industry and General Applications, Vol. IGA-7 (1), January/February 1971.

it is not possible to predict the distribution of the energy deposition.

B. Emergency Procedures

1. Emergency procedures shall be established, distributed, and conspicuously posted near hazardous equipment and followed by personnel during emergencies.

2. Emergency actions include: (a) Evaluating the emergency situation, (b) Turning off the electrical power, (c) Using an appropriate insulator to remove injured personnel from the "live " contact, when necessary, (d) Requesting emergency assistance, and

AN EFFECTIVE ELECTRICAL SAFETY MANAGEMENT PROGRAM

(e) Administering first-aid treatment to an electrical-shock or burn victim as soon as possible including cardiopulmonary resuscitation (CPR) when necessary.

EQUIPMENT SAFETY EXHIBITS (Part VII)

In addition to the general electrical safety design and construction criteria contained in Part III and the general electrical safety operating criteria contained in Part IV, the criteria contained in the following Equipment Safety Exhibits shall also be followed.

The Exhibits cover many of the basic types of electrical equipment used in R&D activities. Each exhibit contains a brief description of the particular equipment; a listing of the types of known hazards; and recommended design, construction, and operating criteria for electrical safety. Because of research specialization, the various DOE laboratories may have special equipment safety requirements that have not been covered; therefore, these Equipment Safety Exhibits are not to be construed as being all inclusive.

EXHIBIT A
CAPACITORS AND CAPACITOR BANKS

1. Description

This exhibit is particularly applicable to capacitors and capacitor banks used as a source of pulsed power, for blocking and filtering, and in oscillator and resonant circuits; where stored energy is generally in excess of 10 joules and terminal-to-terminal or terminal-to-ground voltages generally exceed 300 volts. This exhibit is also applicable to other types of capacitor applications and is intended to complement, not supersede, capacitor installation requirements contained in the current edition of the National Electrical Code.

2. Types of Hazards

a. Capacitors and capacitor banks with stored energy of about 10 joules (or less) can constitute a shock hazard and, with stored energy of about 50 joules (or more), can constitute a lethal shock hazard. The actual degrees of hazards are dependent upon the capacitor or capactor bank voltages in each application.

NOTE: Energy storage in a capacitor is determined by the following equation:

$$W = \tfrac{1}{2}(CV^2)10^{-6}$$

Where W = energy in joules (watt-seconds)
C = capacitance in microfarads
V = potential difference between electrodes, in volts.

b. Because of the phenomenon of "dielectric absorption," all of the charge in a capacitor is not dissipated when it is short-circuited for a short time, and the capacitor can reestablish a substantial fraction of the original voltage after the short circuit is removed. It is also possible for capacitors that are not short-circuited to acquire a charge from local atmospheric electrical disturbances, and by corona from a nearby high-voltage terminal such as on an adjacent, energized capacitor. Thus, any capacitor that is not short-circuited can constitute a safety hazard.

c. A dangerously high voltage can exist across the impedance of a few feet of grounding cable at the moment of contact with a charged capacitor (i.e., when the capacitor is being discharged with the use of a safety grounding hook).

d. Discharging a capacitor by means of a grounding hook can cause an electric arc at the point of contact. Such release of energy can cause burns, due to thermal radiation or flying molten metal.

e. Internal faults may rupture capacitor containers, particularly when parallel-connected capacitors have no individual fuse protection. The force of the explosion may propel pieces of metal, capable of producing serious injury.

f. Rupture of a capacitor (by internal electrical fault) containing combustible dielectric materials, can create a fire hazard.

g. Polychlorinated-biphenyl (PCB) dielectric fluids can release toxic gases when decomposed by fire or the heat of an electric arc.

h. Fuses are generally used to preclude the discharge of energy from a capacitor. If not adequate for this application, such fuses and the capacitors themselves could explode, expelling dangerous projectiles.

i. During transient conditions, capacitors can acquire a charge hazardous to personnel, or an overvoltage harmful to the capacitor itself because of inductance in the circuit in the form of coils or magnets, or in wiring and leakage inductance.

3. Design and Construction Criteria

a. Isolate capacitor banks by elevation, barriers, or enclosures to preclude accidental contact with charged terminals, conductors, cases, or support structures. Design enclosures and barriers to protect personnel from projectiles such as bushings and terminals that might be expelled from the capacitors during a fault.

b. Interlock the circuit breakers or switches used to connect power supplies to capacitors with the capacitor discharge devices. Arrange the interlocking to prevent short-circuiting the power supply when the circuit breakers or switches are closed, and to prevent closing a circuit breaker or switch while the capacitors are short-circuited.

c. Provide capacitors with current-limiting devices, such as fuses and resistors, which are capable of interrupting available fault current or limiting it to safe and manageable values.

d. When fuses are not acceptable for installation in a capacitor bank (e.g., an extremely short discharge time is required), install each capacitor in a suitable enclosure to provide protection to personnel and equipment in case of a fault, or electrically separate the capacitors into sets which are isolated from each other. The number of capacitors in each set shall be such that a faulted capacitor would not receive enough energy from the remaining capacitors in that set to cause it to rupture.

e. Design safety devices, such as shorting switches and grounding switches, buswork, and their associated cables and cable connectors, to withstand the mechanical forces due to the large currents that can result from their operation.

f. Provide bleeder resistors on all capacitors not having discharge devices.

g. Design the discharge-time-constant of currentlimited shorting and grounding devices to be as small as practicable.

h. Provide grounding hooks in sufficient number to assure that protective equipment-grounding requirements are easily satisfied (see Part III.K.).

i. Provide adequate ventilation to keep ambient-air-temperature at capacitor installations at recommended levels.

4. Operating Criteria

a. Restrict access to capacitor areas until all capacitors have been discharged, shorted, and grounded.

b. Remove any residual charge from capacitors by grounding the terminals with low-impedance grounding hooks before beginning to work with them.

c. Do not rely on automatic discharge and grounding devices exclusively for personal safety. Use grounding hooks.

d. Inspect grounding hooks periodically to assure that all connections are secure and that the grounding conductor, and series resistor, if used, are in good condition.

e. Capacitor cases, unless obviously connected to a recognized grounding conductor or grounded structure, shall be considered "charged" and shall be grounded in the same manner as the capacitor terminals in b., above, for personnel safety. Ungrounded capacitor cases should be properly labeled, with their operating voltages identified.

f. Operating personnel should stand clear of cables attached to grounding hooks at the moment of application to a capacitor terminal.

g. Periodically test protective devices, such as automatic shorting switches and grounding hooks, to verify their operation.

h. Inspect capacitor installations routinely for leaking or deformed capacitor cases.

i. Short-circuit all capacitors in storage with a conductor not smaller than #14 AWG, securely fastened to the terminals and left in place until the capacitors are used again. If disposed of, short circuits should also remain in place.

EXHIBIT B
ELECTRICAL CONDUCTORS AND CONNECTORS

1. Description

The conductors and connectors covered in this exhibit are those used in special R&D activities and include high-current (pulsed or continuous), high-voltage, high-frequency, liquid-cooled, and other special conductor and connector applications.

2. Types of Hazards

a. Hazards can occur from conductor insulation damage or deterioration caused by chemical reactions, mechanical forces and abrasion, corona, ionizing radiation, transient over-voltages, or conductor overheating.

b. Metallic cooling-water pipes that are also used as electrical conductors present shock hazards (i.e., they may not be readily reconizable as electrical conductors).

c. Improper application or installation of connectors can result in overheating, arcing, and shock hazards.

d. Inadequate separation between high- and low-voltage cables or terminations can result in hazardous induced voltages and arcing (at the terminations).

e. Dense packing of electrical cables in cable trays or raceways can cause overheating and insulation deterioration (potential fire and electrical arcing hazards).

f. Pulsed operation of cables or high-fault currents can produce sizeable electromagnetic forces. Physical movement of components and exploding cables can create hazardous conditions.

g. Ungrounded or improperly grounded shields of coaxial power cables can result in shock hazards.

3. Design and Construction Criteria

a. Provide conductors that are adequate in current-carrying capacity for their application.

b. Assure that the type of conductor insulation selected is appropriate for the operating and environmental conditions (e.g., thermal ratings; dielectric and mechanical strength; moisture, chemical, and radiation resistance).

c. Provide conductors having capacity ratings sufficiently in excess of the energy supply capability.

d. Where liquid-cooled cables are used, sensing devices (temperature or coolant-flow) should be provided for alarm purposes or equipment shutdown if the cooling-system malfunctions or fails.

e. Derate conductor current capacities commensurate with density of packing (e.g., in cable trays or raceways).

f. Avoid loops (wide spacing) between high-current supply and return conductors to prevent voltage and current induction in adjacent circuits or structural members.

g. Provide suitable routing and additional protection for coaxial cables used in pulsed-power applications where the braid of the coaxial cable rises to high-voltage levels.

h. Provide physical barriers to separate high-voltage conductors from low-voltage conductors, and each other, and design them to withstand fault conditions.

i. Provide adequate labeling, insulation, or other protection for metallic cooling-water piping used as electrical conductors.

j. Provide bracing and conductor supports that are physically and electrically adequate to withstand expected mechanical forces and voltages.

k. Provide connectors that are adequate in current-carrying capacity and voltage rating for their application.

l. Provide appropriate connectors for use with aluminum conductors and assemble them in accordance with approved techniques.

m. Provide adequate separation between adjacent high- and low-voltage cable terminations.

4. Operating Criteria

a. Cables used in recurring experimental activities should be carefully handled and stored between uses.

AN EFFECTIVE ELECTRICAL SAFETY MANAGEMENT PROGRAM

b. Prohibit walking or climbing on cable trays to prevent damage to the cables.

c. Laying cables across the floor for experimental work should be prohibited to the maximum extent practicable. Where this must be done, suitable protection for personnel and cables shall be provided.

d. Cable connectors should be checked and adjusted periodically for tightness, with special attention given to aluminum cable connectors.

e. Plug-in cable connectors, particularly for high voltages or high currents, should be mechanically fastened in place, and the power source shall be deenergized before removing a connector.

EXHIBIT C
ENCLOSURES FOR ELECTRICAL EQUIPMENT

1. Description

This exhibit is particularly applicable to enclosures where equipment voltages exceed 300 volts to ground and where rf radiation or stored-energy electrical components are contained. This exhibit is also applicable to other types of enclosures where the hazards identified herein may also exist.

2. Types of Hazards

a. Electrical shock hazard from ungrounded or poorly grounded enclosures.

b. Burns resulting from rf, eddy current, or microwave heating (with inadequate enclosures).

c. Burns to skin and eyes from electrical arcing and molten metal (with inadequate enclosures).

d. Faults occurring inside the enclosure which may rupture the enclosure and injure personnel or damage adjacent equipment.

e. Failure of interlocks, permitting personnel to come in contact with energized equipment within an enclosure.

f. Crowded working conditions within enclosures, resulting in personnel safety hazards.

3. Design and Construction Criteria

a. Design enclosures so that no contact with live electrical parts can be made from outside, and provide adequate interior working space.

b. Provide electrical interlocks on all doors and easily removable or swinging panels to prevent access without interrupting the interlock circuit. Provide door locks to limit access only to authorized personnel.

c. Provide separate high- and low-voltage and/or instrumentation and control compartments.

d. Provide properly shielded and grounded enclosures for rf power equipment and give particular attention to all openings such as doors, access ports, and viewing windows.

e. Provide enclosures structurally adequate for their intended use. Use adequate material in viewing windows to protect personnel from flying parts that may result from internal electrical faults.

4. Operating Criteria

a. Clear all personnel from an enclosure where hazardous conditions exists before energizing the equipment.

b. On rf systems, use proper instruments at the operating frequency to perform initial and routine measurements of radiation leakage, and take special measurements after equipment modifications or changes in radiation levels.

c. Provide suitable signs and/or warning lights to indicate equipment hazards.

d. When a temporary enclosure is necessary, it should be electrically interlocked, if possible, and should meet the same requirements as a permanent enclosure where practicable.

EXHIBIT D
INDUCTORS, ELECTROMAGNETS, AND COILS

1. Description

This exhibit covers inductors, electromagnets, and coils that are used in the following applications:

a. Energy storage, where power is provided by a dc source or low-frequency ac power supply, and then switched to a load or test area.

b. Inductors used as impedance devices in a pulsed system with capacitors, to provide oscillatory or resonant conditions.

c. Electromagnets and coils that produce magnetic fields to guide or confine charged particles.

d. Inductors used in dc power supplies.

2. Types of Hazards

a. Overheating due to overloads, insufficient cooling, or failure of the cooling system could cause damage to the inductor and possible rupture of the cooling system.

b. Large electromagnets may produce external force fields which can affect the calibration and proper operation of protective instrumentation and controls and can cause nearby motors and transformers to overheat or overload. Such external fields could also attract nearby loose magnetic material and cause injury or damage by impact.

c. Whenever a magnet is suddenly deenergized, production of large eddy currents in adjacent conductive material can cause excessive heating. A fast rate-of-change of field strength, producing high turn-to-turn and terminal voltage can also induce voltages in adjacent conductors which can be hazardous.

d. An inductor is also capable of producing large electromagnetic forces.

e. When one inductor is used with a second, improper conductor polarity can result in abnormal forces and field strengths.

f. Loose and broken inductor or magnet connections can produce excessive heat and arcing.

g. The large amount of energy stored in the field of an energized inductor can damage equipment and injure personnel if it is not discharged in an appropriate manner.

h. Large amounts of stored energy can be released in the event of a "quench" in a superconducting magnet.

i. The relatively long-time constants in highly inductive circuits can cause the continuous release of energy into a fault, producing severe equipment damage and possible fire.

3. Design and Construction Criteria

a. Provide sensing devices (temperature, coolant-flow) which are interlocked with the power source, for safe shutdown of water- or air-cooled inductor and magnet coils in the event of excessive temperatures or cooling-system failure.

b. Where required, fabricate protective enclosures from materials not adversely affected by external electromagnetic fields produced by the equipment.

c. Provide equipment supports and bracing adequate to withstand forces produced during normal operation and fault conditions.

d. Ground electrical supply circuits and magnetic cores wherever feasible and provide adequate fault protection. Provide ground-fault detection for grounded and ungrounded electrical circuits (floating systems), for alarm purposes or equipment shutdown.

e. Provide means for safely dissipating stored energy when excitation is interrupted or a fault occurs.

4. Operating Criteria

a. Provide suitable signs and/or warning lights to indicate equipment hazards.

b. Advise personnel of the hazards of stray magnetic fields by posted instructions or other means. There should be no magnetic material in the clothing or on the bodies of personnel who are in the immediate area of large energized magnets or inductors.

c. Exercise extreme caution when disconnecting the leads of any large inductor. First, lock out the power source and determine that the current has decayed to zero.

d. Exercise caution when checking continuity or measuring resistance of large inductor or magnet coils with a common ohmmeter. Severe shocks can result if both hands are in contact with the terminals when the test probes are removed.

EXHIBIT E
INSTRUMENTATION AND CONTROL SYSTEMS

1. **Description**

This exhibit covers instrumentation and control systems that are associated with R&D equipment and operation.

2. **Types of Hazards**

   a. Instrumentation and control systems may involve circuits which operate at hazardous voltage levels (e.g., 115 volts) and/or may be served from high-current power supplies.

   b. Failure of insulating and/or isolating devices could bring instrumentation and control systems in contact with circuits or components operating at hazardous voltages or currents.

   c. Failure or malfunction of instrumentation can produce erroneous readings and prevent recognition of hazardous conditions.

   d. Failure of control circuits can cause unintentional operation of hazardous equipment and/or inhibit the operation of safety devices such as enclosure interlocks, warning lights, and overload protection.

3. **Design and Construction Criteria**

   a. Provide electrical instrumentation and control circuitry with adequate isolation at its interface with the main power equipment being controlled and monitored. Consider both the normal and the fault conditions which can exist during operation of the main equipment. This includes physical separation of high- and low-voltage circuitry and equipment and use of surge protectors and isolation devices such as transformers, high impedances, optical coupling, or telemetering.

   b. Assure that relay and interlock contacts on instrumentation and protective circuits are rated at least as high as the voltage of the circuit and that current ratings are as high as the normal disconnect rating of the protective fuse or circuit breaker used. Give particular attention to the inductance of the circuit in the proper application of relays and interlocks.

c. Design control circuits as fail-safe as possible so that loss of power does not result in a hazardous operating condition.

d. Provide redundant controls and instrumentation on sections of a system where a single failure could otherwise result in a hazardous operating system.

e. Provide a clear indication of the status of hazardous remotely controlled equipment with positive feedback for each specific command.

f. Provide shorting devices for use on current transformers when instruments or controls are connected or disconnected.

g. Properly label or cover exposed high-voltage terminals where higher voltages (e.g., 115 volts) are present in low-voltage (e.g., 30 volts) instrumentation and control compartments.

h. Provide separate raceways or physical isolation between instrumentation and high-current conductors.

i. Wire control circuits so that "sneak" circuits cannot exist, and accidental grounding of one conductor cannot cause safety devices to become inoperative.

j. Use consistent labeling for control panels and provide graphic control displays for large systems.

k. Route control wires such that no large-looped circuits are formed.

l. Design protective interlock circuits and power-supply controls so that reactivation of the interlock circuit (i.e., completing the circuit will not result in automatic restoration of power to the equipment.

4. Operating Criteria

a. Carefully inspect and test all new or modified instrumentation and control systems to assure that they perform in accordance with operating and safety requirements. Also include simulation of failures, operation of upper- or lower-limit control features, safety interlocks, and interlock systems.

AN EFFECTIVE ELECTRICAL SAFETY MANAGEMENT PROGRAM 251

b. Give immediate attention to malfunctions or failures of instrumentation and control systems adversely affecting safety, and take corrective action.

c. Test safety interlocks at least annually to assure operability. The use of written check lists is recommended.

d. Bypass electrical interlocks and other safety devices only when absolutely necessary. Establish procedures for tagging the interlock and logging its location and the time when bypassed and restored; also, provide for written approvals prior to bypassing an interlock.

EXHIBIT F
RADIO-FREQUENCY EQUIPMENT

1. Description

Radio-frequency (rf) equipment covered by this exhibit is that class of apparatus used for pulsed or continuous generation of high-frequency energy and microwave radiation. For the purpose of this exhibit, high-frequency energy refers to that portion of the electromagnetic spectrum corresponding to frequencies •1,000 MHz. Microwave radiation is defined as electromagnetic radiation having frequencies of 1,000-30,000 MHz, or wavelengths of 30-1 cm, respectively. The Occupational Safety and Health Act (OSHA) 1972 Regulations have adopted a Radiation Protection Guide for the 10 MHz-10 GHz frequency range of 10 milliwatts-per-square-centimeter (mW/cm^2) averaged over any possible 6-minute period. Regulations of the Bureau of Radiological Health (HEW) governing microwave-oven leakage provide for a limit of 1 mW/cm^2 at a distance of 5 cm from the surface of a new oven and a limit of 5 mW/cm throughout its useful life.

2. Types of Hazards

a. Until recently, studies of the biological effects of electromagnetic radiation have been directed primarily toward X and gamma radiation. The growing use of microwave radiation has stimulated a great deal of interest in possible effects of nonionizing radiation. However, quantitative data on biological effects for the microwave-frequency range are inconclusive.

b. Biological effects produced by microwaves are caused primarily by the heating of body tissue (hyperthermia). Organs lacking adequate blood supply for needed temperature regulation are the most susceptible. Biological effects are most pronounced in abdominal organs (tissue damage), eyes (cataracts), and testicles (tissue damage).

c. Severe rf burns may result if the body comes in contact with, or even near, a source of high-frequency energy such as an induction heating load coil. Metal objects (e.g., watchbands, rings, or pens) can be heated by rf energy under these conditions and cause burns.

d. Electromagnetic radiation from rf equipment may induce energy in other apparatus which can interfere with the operation of associated electrical circuits, including control circuits and bioelectronic implants such as heart pacemakers.

e. Radiofrequency (rf) energy may be reflected by a variety of materials which can result in injuries.

3. Design and Construction Criteria

a. Provide shielding and other control measures to minimize radiation leakage.

b. Make provisions for excluding unauthorized personnel from rf equipment areas by using caution signs and other visible and/or audible signals. Include the OSHA-approved warning symbol for nonionizing radiation (see Article 1910.97).

c. Guard exposed rf load coils appropriately with insulating material to prevent accidental contact by operating personnel.

d. Provide an adequate external ground for rf equipment to dissipate stray rf energy.

e. Avoid sharp edges or points which might emit corona discharges.

f. Design high-frequency heating equipment with adequate clearances for rf leads and lengths of nonconducting cooling-water hose.

g. Use rf bypassing (capacitors or chokes) on control power in instrument leads which enter the rf compartment.

h. Size the viewing ports to be small in comparison to the wavelength of the rf energy or otherwise shield them.

4. Operating Criteria

a. Wear suitable protective goggles in exposed areas. Goggles with an evaporated gold firm are the most practical type for microwave radiation.

b. Eye examinations shall be given to personnel prior to their assignment to operations involving microwave facilities, and re-examinations shall be given periodically.

c. Establish and implement procedures for minimizing personnel exposure to rf energy.

d. provide area-monitoring equipment with sensing heads that are nondirectional in their pickup patterns and consider the use of pocket-type microwave dosimeters.

e. Prohibit personnel from making close visual examination of energized microwave radiators, reflectors, etc.

f. During test operations, use dummy loads such as water or other absorbent material.

g. Establish and implement procedures for frequent periodic monitoring (scanning) of interface couplings to detect microwave energy leaks.

h. Make sure that personnel working with or near high-frequency power sources are not wearing metallic objects (e.g., watch bands or rings).

i. When test procedures require free space radiation, position the radiation device where the energy beam is not directed toward personnel. When a device is positioned, the probable directions of reflected beams should also be considered.

j. Avoid the use of electrical equipment near induction heating apparatus to prevent induced energy from interfering with thge equipment's operation.

k. Post safety instruction, including the hazards and possible consequences of overexposure to rf energy, where applicable.

EXHIBIT G
POWER SUPPLIES

1. Description

This exhibit is particularly applicable to high voltage (over 600 volts terminal-to-terminal or 300 volts-to-ground) ac or dc power supplies and to low-voltage, high-current ac or dc power supplies, used in special R&D activities. This exhibit is also applicable to other types of power supplies where the hazards identified herein may also exist.

2. Types of Hazards

a. It is possible that a power supply in a remote location could be energized and personnel could unknowingly come in contact with the energized equipment (connected load).

b. Electrical faults or switching transients can cause voltage surges in excess of the normal terminal voltage rating of the power supply.

c. Interal component failure can cause excessive voltages on external metering circuits and low-voltage auxiliary control circuits.

d. Overload or improper cooling can cause excessive temperature rise, resulting in possible equipment damage and associated hazards.

e. Electrical faults can cause conductors to melt and cause other components, such as insulating materials, to melt, explode, or burn.

f. Output circuits and components can remain energized while input power is interrupted due to parallel power sources or stored energy in reactive components (e.g., capacitors).

g. Auxiliary and control power circuits can remain energized when the main power circuit is interrupted.

h. When power supplies serve more than one experiment, switching errors can result in energizing the wrong equipment (load), creating hazards to nearby personnel.

i. Overcurrent protective devices such as fuses and circuit breakers for conventional applications may not adequately limit or interrupt the total inductive energy and fault currents in highly inductive dc systems.

3. Design and Construction Criteria

a. Do not install circuits, components, or other equipment not essential to the power supply within the power-supply enclosure.

b. Provide isolation devices or physical barriers to prevent high-voltage stored energy from being dissipated in a low-voltage supply and/or control circuits. Two means of isolation shall be considered when the power supply is associated with a high-energy system to assure sufficient reliability so the failure of one device does not result in personnel injury or excessive equipment damage.

c. Provide an automatic switch and/or fixed-bleeder resistor in the output circuit for discharging the power supply when the input power is turned off.

d. Provide overcurrent, undervoltage, or other protection for both power supply and load, as appropriate.

e. For each power supply, clearly identify the main power input circuit breaker (disconnect); locate it within sight of the power supply, if feasible; and equip it with lockout provisions. In addition, install a second means of input power shutdown directly at the power supply.

f. Provide alarms, signs, or lights to warn personnel that the power supply is energized, especially on remote loads.

g. Minimize the number of control stations and provide emergency stop controls at all remote power supply locations, with lockout provisions when deemed necessary.

h. Avoid multiple-input power sources.

i. For inductive loads, connect thyrite or equivalent devices across the power-supply dc terminals to assure satisfactory discharge of stored energy.

j. Determine that fault limiting or interrupting devices, such as fuses or circuit breakers, are proper for the system in which the power supplies are used.

4. Operating Criteria

a. Prior to initial operation and periodically thereafter, carefully inspect the power supply and calibrate and check all protective devices.

b. Before entering power-supply or associated equipment enclosure, take the following precautions:

1. Deenergize the equipment.
2. Open and lock out the main input-power circuit breaker.
3. Check for auxiliary power circuits which could still be energized.
4. Inspect automatic shorting devices to verify proper operation.
5. Short the power supply from terminal-to-terminal and terminal-to-ground with grounding hooks.

c. Label equipment to identify input power sources; and label input power sources to identify their connected power-supply loads.

d. Equipment that is remotely controlled or unattended while energized should be labeled with emergency shutdown instructions and identification of responsible personnel to contact in case of emergency.

EXHIBIT H
RESISTORS AND RESISTOR BANKS

1. Description

This exhibit covers resistors and resistor banks in the following R&D applications:

a. To carry pulsed current exceeding their steady state ratings.

b. To perform safety functions, such as grounding.

c. To absorb the discharge of stored energy.

d. To provide a means of connecting instruments to a high-voltage circuit, as in a voltage divider.

2. Types of Hazards

a. Resistors used in discharge circuits can be damaged when operated a high-current or high-voltage levels.

b. Resistors used in pulsed circuits can be subjected to overvoltages which can cause electrical arcs, with possible damage or failure.

c. Failure of a resistor used in a discharge device for an energy storage system can create a hazardous condition if the discharge circuit does not function as intended during subsequent operation of equipment.

d. Large currents, due to faults or abnormal circuit operation, may produce forces capable of destroying resistors. Damage to adjacent equipment and injury to personnel may result.

e. In the voltage divider of a high-voltage metering circuit, failure of a resistor used in the low-voltage section can create hazardous voltages on the low voltage metering circuit and at the instrumentation location.

f. Resistors requiring liquid or forced-air cooling media are subject to overheating and failure if the cooling system fails.

g. Hazardous voltages can develop across resistors in grounding circuits when fault or discharge current flows.

h. Resistors may operate at temperatures high enough to cause severe burns to personnel or ignite combustible materials.

i. Failure of an inductor discharge resistor, such as that used for a motor field winding, can result in hazardous and destructive voltages in the motor circuit.

j. Failure of a power-supply bleeder resistor could expose personnel to hazardous voltages.

k. Electrolytic resistors may explode on overvoltage and may open-circuit if the fluid level gets low.

3. Design and Construction Criteria

a. Select resistors for pulsed operation so that any possible suuccession of pulses in the circuit will not raise the temperature of the resistors to levels harmful to the resistors themselves or to surrounding equipment.

b. Provide resistors that are capable or withstanding any transient overvoltages to which they may be subjected.

c. Resistors used where large pulse or fault currents may be expected should be strong enough to withstand the resulting magnetic forces. They should be installed in an enclosure capable of minimizing damage and preventing injury if a failure should occur.

d. When failure of a resistor could expose personnel to hazardous voltages, consider the installation of two or more resistors in parallel, each rated for maximum operating conditions.

e. Install temperature of flow-sensing devices in resistor installations which require external air or liquid cooling.

f. Resistors operating at voltages or temperatures hazardous to personnel should be installed in an enclosure with the access interlocked to prevent entry while the resistors are energized.

g. Provide resistor enclosures that are well ventilated and constructed of noncombustible material.

h. Provide adequate signs and/or warning lights to warn personnel of the hazards present in resistor installations.

i. Protect resistors used in high-voltage circuits from surface contamination by adverse environmental conditions.

j. Assure that the insulation of conductors used to connect resistors is adequate for the temperatures and voltages to be encountered.

k. Install resistors in a manner which precludes damage to adjacent components due to heat.

l. Design resistor networks and configurations so that each resistor operates within its rating or capability.

4. Operating Criteria

a. Periodically inspect the condition of resistors and resistor banks including resistor connections.

b. Test resistor enclosure safety-interlock systems at least annually.

EXHIBIT I
ELECTRICAL SWITCHES

1. Description

This exhibit covers special electrical switches used in R&D activities where safety requirements are not specifically covered by existing codes.

2. Types of Hazards

a. Electrical shock from contact with exposed, energized switch parts is a common hazard.

b. Sufficient energy may be developed under fault conditions to cause the switch to explode.

c. Arcing at switch terminals under transient or fault conditions can subject the isolated section of a circuit to hazardous volatges and power levels (i.e., flashover across open switch contacts).

d. Electrically controlled switches operated unintentionally, due to malfunction of the control circuits, present shock hazards.

e. Switches opened under load conditions can create severe arcing.

f. Switches used above their voltage and current ratings present shock and burn hazards.

3. Design and Construction Criteria

a. Provide suitable switches such that under fully loaded and fault conditions their voltage, current, and interrupting ratings are not exceeded.

b. Provide a positive indication of switch positions.

c. Provide a system of interlocks which interrupts the normal operating control power to the switch during testing or maintenance periods.

d. Provide locking features on switches to prevent operation when personnel are working on connected circuits or equipment.

e. Provide lockable switch enclosures for access control purposes.

f. Assure that switches not designed to disconnect under load conditions cannot be opened when the circuit is energized.

g. Provide protective covers and/or barriers, wherever practicable, to prevent personnel contact with live parts.

h. Provide suitable means for locking disconnect and/or manual transfer-type switches in the desired position. Wherever possible, however, switches should *not* be locked in the *ON* position, preventing load-side circuits from being readily deenergized.

4. Operating Criteria

a. Utilize adequate lock-and-tag procedures before working on connected circuits or equipment.

b. Establish and implement operating procedures for checking that no one is working on the load and that all protective grounds have been removed before restoring power.

c. Periodically inspect switch conditions and perform operating tests.

EXHIBIT J
STORAGE BATTERIES AND BATTERY BANKS

1. Scope

This exhibit covers recharageable-type batteries used for storage of electrical energy. These criteria are not limited to batteries of a particular voltage and energy rating, since the nature of the associated electrical hazards is similar for any battery size except that the severity of the hazard increases with increased battery rating.

2. Types of Hazards

a. Accidental grounding of one polarity of a battery bank can create a hazardous voltage between the ungrounded polarity and ground.

b. Accidental shorting of the exposed terminals or cables of a battery can result in severe electric arcing, causing burns and electrical shock to nearby personnel.

c. Hydrogen gas generated during battery charging can create fire and explosion hazards.

d. Exposed terminals of a battery bank present electrical shock hazards.

e. Batteries may explode if they are shorted or charged at excessively high rates, particularly so for sealed-cell batteries.

f. Electrolytes may be highly corrosive and can produce severe burns to personnel on contact.

3. Design and Construction Criteria

a. Battery installations shall conform to requirements in the latest edition of the National Electrical Code.

b. Battery banks should not be grounded. A ground detector should be used to indicate an accidental ground.

c. Mount batteries so as to allow safe and convenient access for maintenance.

d. Provide lockable doors to rooms or enclosures containing battery banks for access control purposes.

e. Deluge safety showers and eye-wash stations should be provided in close proximity to battery banks.

4. Operating Criteria

a. Maintain battery-bank connections (cell-to-cell and terminal) clean and tight to prevent excessive heating due to contact resistance.

b. Do not repair battery connections when current is flowing; an accidental opening of the circuit could result in a hazardous arcing condition.

c. Clearly post electrical and other hazards of battery banks and emergency first aid procedures near the equipment.

ADDITIONAL CHEMICAL, FIRE, AND OTHER HAZARDS ASSOCIATED WITH OPERATION OF ELECTRICAL R&D EQUIPMENT (PART VIII)

A. Introduction

The operation of electrical apparatus can result in indirect hazards and, while not electrical in nature, are intimately associated with the equipment. These hazards may involve physiological and toxic effects, fire, explosion, corrosion, failure of safety systems due to nonelectrical causes, and many others. This part is not intended to present detailed criteria for each; rather, it is a listing of some types of hazards which could be encountered, some general design and construction criteria, and operating criteria which need to be followed when working with electrical equipment where such hazards may be present. Appendix B lists sources for standards, criteria, and guides that cover safety aspects in greater detail for many of these hazards.

B. Types of Hazards

1. *Ozone.* Many electrical devices generate significant quantities of ozone due to sparking, corona, or ultraviolet light. Ozone, in concentrations as low as 1ppm, can cause adverse physiological effects.

2. *Hydrogen.* Hydrogen is often used for bubble chambers, accelerator targets, and cryogenic magnets, and is also a byproduct of battery-charging or other electrolytic-type operations. It is a highly flammable, highly explosive gas with a lower explosive limit of about 4% in air. Usually, there is some associated electrical equipment which may act as an ignition source during a hydrogen release.

3. *Superconducting Devices.* The increasing use of superconducting magnets and other devices presents several hazards. If hydrogen is the cryogenic fluid, there are potential fire and explosion hazards. Cryogenic temperatures present severe burn hazards. When a cryogenic magnet quenches, it heats up and can cause a sudden pressure build-up as the liquid turns to gas. This pressure build-up can rupture the containment vessel and create an explosion-type hazard.

4. *Polychlorinated-Biphenyl (PCB) Oils.* Oil which have been modified with the use of PCB to be fire-resistant present hazards to both personnel and equipment. These oils are toxic when ingested or absorbed through the skin. The presence of water in the oil causes hydrochloric acid formation which can damage equipment and result in equipment failures. When decomposed by heat, these oils release toxic gases such as phosgene.

5. *Batteries.* In addition to the electrical hazards associated with storage batteries, there are other major hazards.

a. Batteries having a sulfuric acid electrolyte present severe burn hazards.

b. During charging, hydrogen is generated in the cells and an inadvertent spark can cause the battery to explode, and other potentially dangerous gases (e.g., stibine and arsine) may also be generated.

6. *Noise Sources.* Continued exposure to very high noise levels, such as may be present in the vicinity of some electrical equipment, can cause deterioration in hearing ability. Sudden loud noises, such as a spark gap firing or capacitor bank discharging, can create a safety hazard by startling personnel who might be working nearby on hazardous equipment.

7. *Coolants.* The coolants used in electrical equipment are most commonly water, oil, and an antifreeze liquid such as ethylene glycol, and their release can be hazardous. Water can create a leakage path for voltage discharge, initiate short circuits, interfere with interlock systems, or create low-resistance ground paths. Oil and glycol when released present a slip hazard which can result in physical injury.

8. *Environment.* Any device used as part of a safety system may be subject to malfunction as a result of environmental conditions. The environmental factors to be considered include temperature, dirt and dust, moisture, ice, dc magnetic fields, and other electromagnetic fields. A relay can fail to operate in the presence of a large stray magnetic field; dirt or dust can make an electrical contact inoperative; and freezing of moisture can prevent an interlock switch from operating. Such failures may make a safety system inoperative.

9. *Fire.* There are inherent ignition sources present in most electrical equipment and fire is always a potential danger. Many of the materials used in construction of electrical equipment are combustible. The commonly used insulating materials can generate large amounts of smoke, with toxic fumes resulting from the chlorine content.

10. *Thermal Sources.* Electrical equipment may contain devices operating at temperatures which can cause thermal burns or initiate fires.

11. *Moving Mechanical Devices.* Unprotected actuators, fans, blowers, gears, pulleys, etc., present physical safety hazards.

12. *Light Sources.* Lasers, ultraviolet and bright-white light sources, spark gaps, and other devices present severe eye hazards. Ultraviolet light, even during the short exposures, can cause conjunctivitis which could lead to permanent eye damage. Laser beams can cause retinal damage and severe burns to exposed body areas.

13. *Magnetic Fields.* There is concern about effects to biological organisms due to exposure to dc magnetic fields. Unwarranted long-term exposures, e.g., near large bubble chambers, should be minimized. Carrying magnetic tools or equipment near magnets with large stray fields could lead to physical injuries.

14. *Electromagnetic Radiation.* High-power, pulsed electromagnetic radiation can cause secondary effects near the source, such as sparking between any conducting materials in the area and heating of nearby objects, and resultant personnel hazards (e.g., body burns).

15. *Bioelectronic Implants.* Implants such as heart pacemakers may be sensitive to electromagnetic fields. Persons with such implants should avoid exposure to this potential hazard.

16. *X Rays.* Many high-voltage devices generate x rays as an undesirable by-product. Such devices include klystrons, high-voltage rectifiers, and high-voltage vacuum tubes. Shielding requirements should be given proper attention.

17. *Nuclear Radiation.* Electrical and electronic equipment adjacent to reactors and particle accelerators may be subject to nuclear radiation, resulting in induced radioactivity. Before any work is performed on such equipment, it should be monitored for radioactivity and appropriate precautions should be taken to minimize exposure to personnel.

18. *Energy Storage Devices.* In addition to their electrical hazard, energy storage devices have the capability of creating severe arcs and fireballs involving vaporization and

scattering of copper, steel, or other materials in the arc vicinity. Severe skin, flesh, and eye burns can result from the initiation of such an arc. An exploding capacitor can project bushings and other material a considerable distance. Strong magnetic forces in coils and magnets can cause structural failure.

C. Design and Construction Criteria

1. Provide adequate ventilation in areas where any toxic fumes may be present.

2. Provide eyewash and safety showers near battery banks or other equipment containing acids or other hazardous fluids.

3. Eliminate ignition sources in the vicinity of any equipment using or generating hydrogen.

4. Provide adequate shielding for x ray sources and for high voltage devices which produce x rays as a by-product.

5. Provide proper fire detection and protection equipment when ignition sources exist in the presence of significant amounts of combustible material.

6. Provide proper environmental protection for safety devices.

7. Provide proper drainage for coolant releases.

8. Provide adequate protection against thermal burns for hot equipment.

9. Provide adequate safety guards on moving gears, belts, pulleys, and other mechanically moving devices.

10. Provide adequate shielding for high-power rf sources.

11. Provide suitable area-monitoring equipment to warn of hazardous situations.

D. Operating Criteria

1. Limit access to hazardous areas to only qualified personnel by means of barriers, key interlocks, padlocks, etc.

2. Provide warning signs indicating the presence of any potential hazard such as toxic fumes, laser light, ultraviolet light, noise, toxic chemicals, nuclear radiation, fire hazards,

magnetic fields, electromagnetic radiation, and hydrogen or other explosive gases. These signs should conform to OSHA guidelines whenever possible.

3. Post emergency procedures to be followed in case of accidental exposure to hazards.

4. Provide proper safety equipment such as laser safety goggles, protective clothing, breathing apparatus, safety glasses, and safety helmets.

APPENDIX A
DEFINITION OF TERMS

The following definitions apply to terms used in these Electrical Safety Criteria.

Shall   Denotes a requirement.

Should   Denotes a recommendation.

Can   Denotes a possibility.

May   Denotes permission (neither a requirement nor a recommendation).

Knowledgeable Person   One who is recognized by laboratory management as having sufficient understanding of an experimental device or facility to be able to positively identify and control the hazards it may present.

Companion   Co-worker who is cognizant of the danger and occasionally checks on the other worker.

Safety Watch   A person whose specific duties are to observe the worker(s) and operations being performed, prevent careless acts, quickly deenergize the equipment in emergencies and alert emergency rescue personnel. This person shall have basic cardiopulmonary resuscitation (CPR) training, and be thoroughly instructed on the locations of emergency shut-off switches and power disconnects and on the specific working procedures to be followed.

Massive Ground   Large areas of grounded metal surfaces on, or adjacent to equipment (e.g., equipment supports, enclosures and floor grating) and earth, in the case of outdoor equipment, that make isolation difficult or impossible for the protection of personnel working on energized electrical equipment.

Electrical Hazard A potential source of personnel injury, resulting either directly or indirectly from the use of an electrical energy source.

  a. *Direct Electrical Hazard* A potntial source of personnel injury resulting from the flow of electrical energy through a person's body (potential electrical shock and burns).

  b. *Indirect Electrical Hazard* A potential source of personnel injury resulting from electrical energy transformed into other forms of energy (e.g., radiant energy such as rf energy, light and heat; energetic particles; mechanical forces; and chemical reactions, such as fire and the production of noxious gases and compounds).

First-Line Protection The primary protective system and/or operating procedures, provided to prevent physical contact with energized equipment by personnel.

Back-Up Protection A secondary, redundant, protective system provided to deenergize a facility to permit safe physical contact by personnel. This system shall be totally independent of the first-line protection and capable of functioning in the event of total failure of the first-line protective system.

Fail-Safe Built-in safety characteristics of a unit or system such that failure (of unit or system) or loss of control power will not result in an unsafe condition.

Grounding Points Most direct connections possible for protective grounding of a source of stored energy (e.g., capacitor terminals).

Positively-Deenergized All external sources of electrical energy are disconnected by some positive action (e.g., locked-out switch or circuit breaker) *and* all internal energy sources (e.g., capacitor energy storage) are made safe by discharging and grounding.

Lock-and-Tag Procedures Written procedures covering: (a) the use of padlocks or other key locks to assure that energy-supply disconnect devices are maintained in the "open" position for safe work on equipment, (b) the use of easily recognizable "tags" attached to the disconnect device control that identify pertinent information on the work being performed, and (e) other information necessary to assure the safety of personnel and protection of equipment, both during performance of the work and on restoration of energy supply following work completion.

Grounding Hook A device for making temporary connection to dicharge and ground internal energy sources in hazardous electrical equipment, consisting of a bare copper rod formed in the shape of a "shephers's hook" at one end, equipped with an insulating handle, securely connected at the other end to a suitable bare flexible copper cable which, in turn, is clearly visible through its insulating sheath and securely connected to equipment or building ground.

APPENDIX B
BIBLIOGRAPHY

1. *National Electrical Code.* National Fire Protection Association, Boston, 1978, NFPA No. 70-1978; also ANSI C2.
2. *National Electrical Safety Code,* 1977 Ed., National Bureau of Standards. Institute for Electrical and Electronic Engineers, Inc., New York, 1977; also ANSI C2.
3. IEEE Standard 142-1972. Recommended Practice for Grounding of Industrial and Commercial Power Systems. Institute of Electrical and Electronic Engineers, Inc., New York, 1972.
4. IEEE Standard 141-1976. Electric Power Distribution for Industrial Plants. Institute of Electrical and Electronic Engineers, Inc., New York, 1976.
5. ISA Monograph 112. Electrical Safety Practices. Instrument Society of America, Pittsburgh, 1968.
6. HEW, *Particle Accelerator Safety Manual.* U.S. Department of Health, Education, and Welfare, National Center for Radiological Health, Rockville, Md., 1968, MORP 68-12.
7. ANSI Standard C55.1-1968. Shunt Power Capacitors. American National Standards Institute, Inc., New York, 1968.

8. ANSI Standard C39.5-1974. Safety Requirements for Electrical and Electronic Measuring and Controlling Instrumentation. American National Standards Institute, Inc., New York, 1974.
9. R. J. Vetter, P. L. Ziemer, and D. Puntenny, Microwave Dosimetry. *Res./Dev.* 25, 22-24 (April 1974).
10. Stabilized Power Supplies Direct-Current Output. National Electrical Manufacturers Association, New York, Publication PY-1972.
11. Motors and Generators. National Electrical Manufacturers Association, New York, Publication MG1-1972.
12. ANSI Standards C50. Rotating Electrical Machinery. American National Standards Institute, Inc., New York.
13. IEEE Standard 484-1975. Recommended Practice for Installation of Large Lead Storage Batteries for Generating Stations and Substations. Institute of Electrical and Electronic Engineers, Inc., New York, 1975.
14. N. Irving Sax. *Dangerous Properties of Industrial Materials*, 4th Edition. Reinhold Publishing Co., New York, 1975.
15. Fire Protection Guide on Hazardous Materials, 6th Edition. National Fire Protection Association, Boston, 1978. *NFC 11*, 704 (1978).
16. ANSI Standard Z136.1-1976. American National Standards for the Safe Use of Lasers. American National Standards Institute, Inc., New York, 1976.
17. *Accident Prevention Manual for Industrial Operations*, 7th Edition. National Safety Council, Chicago, 1974.
18. Threshold Limit Values of Airborne Contaminants. American Conference of Governmental Industrial Hygienists, Sec/Treas, P. O. Box 1937, Cincinnati, OH, April 1977.
19. OSHA 29CFR, Part 1910. General Industry Safety and Health Regulations. Occupational Safety and Health Administration, U.S. Department of Labor, Washington, D C, 20210.

20. R. H. Lee. Electrical Safety in Industrial Plants. *IEEE Trans. Ind. Gen. Appl., IGA,* 7 (1) January/Feburary 1971.
21. National Fire Codes, National Fire Protection Association.

APPENDIX C
ELECTRICAL SAFETY CRITERIA COMMITTEE MEMBERSHIP

Wayne H. Gardner
 (Committee Chairman)
Senior Engineer
Office of Construction and
 Facility Management
Department of Energy

Carlo Ferraro, Jr.
 (Committee Coordinator)*
Safety Engineer
Division of Operational and
 Environmental Safety
Department of Energy

Delwyn D. Bluhm
Section Head,
 Project Engineering
Engineering Services Group
Ames (Iowa) Laboratory

John F. Burdoin†
Electrical Engineer
 Security Systems and Hazards
 Control Group
Electronics Department
Lawrence Livermore
 Laboratory

Maynard Cowan, Jr.
Manager, Simulation
 Research Dept.
Physical Research
Sandia Laboratories

Donald A. Davis
Power Engineer
Accelerator Department
Brookhaven National
 Laboratory

Edward C. Hartwig
Department Head
Electronics Engineering
Lawrence Berkeley Laboratory

Henry Kacinskas
Electrical Engineer
Engineering Division
Argonne National Laboratory

Sigmund Mosko
Electrical Engineer and
 Cyclotron Operations
 Manager
Physics Division
Oak Ridge National
 Laboratory

John G. Murray
Senior Professional Technical
 Staff Member
Tokamak Fusion Test
 Reactor (TFTR) Engineering
Princeton Plasma Physics
 Laboratory

Martin Plotkin
Senior Engineer
Intersecting Storage Accelera-
 tor (ISA) Project (ISABELLE)
Brookhaven National
 Laboratory

Thomas M. Putnam
Assistant Division Leader
 for Safety
Medium Energy Physics Division
Los Alamos Scientific
 Laboratory

Alexander Tseng
Chief Electrical Engineer
Plant Engineering
 Technical Division

Stanford Linear
 Accelerator Center

Vincent P. Zernoski‡
Electrical Safety
 Engineer
FERMI National
 Accelerator Laboratory

Present affiliations:

*Head, Explosives and Nuclear Weapons Safety Section, Office of Chief of Naval Operations, Department of Navy.

†Electrical Engineer, Plant Systems, Nuclear Regulatory Commission.

‡Electrical Hazards Control Engineer, Princeton Plasma Physics Laboratory.

Index

A

Accident statistics, 223
Accidents, case histories, 199-201
American National Standards Institute, 2, 3
American Wire Gage size, 19

B

Burns from electric shock, 11

C

Cardiac arrest, 18
Code of Federal Regulations, 3
Construction electrical hazards, 36, 37
  installation requirements, 37
  [Construction electric hazards] examination, installation, and use, 38
  grounding, 39
  guarding, 38
  overcurrent protection, 39
  safety-related maintenance equipment, 41
  environmental deterioration, 41
  safety-related work practices, 40
  lockout and tagging, 40, 41
  passageways and open spaces, 40
  protection of employees, 40
  special equipment, 41
  batteries, 41
  battery charging, 42
Construction electrical standards, OSHA 3097, 36
Current interruption, 9

D

Defibrillation, 11

E

Edison, Thomas, 2
Effects of electrical currents on humans, 14, 15
Electric shock causes, 17, 18
Electrical accident statistics, 17
Electrical current effects on human body, 6
Electrical hazards and emergency procedures, 236
  electrical hazards, 236, 237
  emergency procedures, 238
Electrical hazard protection, 18
Electrical safety administration, general criteria, 225
  administrative policy, 226
  codes, standards, and criteria, 227
  safety organization, 226
Electrical shock, 5
Electrical workers survey, 201
  general findings, 210
  investigative approach, 202, 203
  questionnaire data, 204-210
    communications problems, 212, 213
    comparison of management and nonmanagement responses, 216
    conflicting pressures, 214
    lack of electrical standards, 215

[Electrical workers survey]
    policy, 211
    reporting of incidents and accidents, 216
    training programs, 215
  recommendations, 217
    communications and controls, 219
    electrical standards, 220, 221
    policy, 217, 218
    training, 221
Electricution, 1, 6, 10, 11
Energized equipment safety rules, 32, 33, 34, 35
Equipment safety exhibits, 239
  bibliography, 268-269
  capacitors and capacitor banks, 239
    design and construction criteria, 241, 242
    types of hazards, 239, 240
  chemical, fire, and other associated hazards, 261
    design and construction criteria, 265
    operating criteria, 265
    types of hazards, 262-265
  definition of terms, 266-268
  electrical conductors and connectors, 243
    types of hazards, 243, 244
  electrical switches, 258
    design and construction criteria, 259
    operating criteria, 260
    types of hazards, 259
  enclosures for electrical equipment, 245
    design and construction criteria, 246

INDEX

[Equipment safety exhibits]
  operating criteria, 246
  types of hazards, 245
inductors, electromagnets,
    and coils, 246
  design and construction
    criteria, 248
  operating criteria, 248
  types of hazards, 247
instrument and control systems, 249
  design and construction
    criteria, 249, 250
  operating criteria, 250
  types of hazards, 249
power supplies, 254
  design and construction
    criteria, 255
  operating criteria, 255, 256
  types of hazards, 254
R&D safety committee, 270
radio-frequency equipment, 251
  description, 251
  design and construction
    criteria, 252
  operating criteria, 253
  types of hazards, 251
resistors and resistor
    banks, 256
  description, 256
  design and construction
    criteria, 257, 258
  types of hazards, 256, 257
storage batteries and battery
    banks, 260
  design and construction
    criteria, 261
  operating criteria, 261
  scope, 260
  types of hazards, 260

F

Fatal accidents, 5
First aid methods, 13
Franklin, Benjamin, 1, 2

G

Grounding, definition, explanation, 20
Guarding against accidental contact, 19

H

Heat sensations, 7
High current effects, 10, 11
High-frequency effects, 6
Home electrical hazards, 42
  extension cords, 44
  fuse boxes, 43
  switches, 44
  wall electrical outlets, 43
Human contact surfaces, 7
Human effects of electrical accidents, 5

I

Insulation effects, 19
Institute of Electronic and Electrical Engineers, 3

L

Let-go currents, d.c., 8
Let-go currents, 60-cycle, 7, 9
Let-go voltages, 60-cycle, 9

M

Man vs. woman experiments, 7, 8
Management of effective electrical safety program, 223
Mechanical devices for control of hazards, 21
Muscular contraction, 1, 7
Muscular reactions, 9

N

National Association of Fire Engineers, 2
National electric codes, 2
National Electrical Code 1987, 47
  communications systems, 61
  equipment for general use, 55
  requirements for electrical installations, 50
  scope, 48
  special conditions, 60
  special equipment, 58
  special occupancies, 57
  tables and examples, 63
  wiring design and protection, 54
  wiring methods and materials, 54
National Electrical Safety Code, 65, 66
  employees, 108-110
  exits, 81
  fire extinguishing equipment, 82

[National Electrical Safety Code]
  floor, floor openings, passageways, stairs, 80
  grounding conductor and means of connection, 73
    ampacity and strength, 74
    bonding of equipment frames and enclosures, 75
    composition of grounding conductors, 73
    connecting of grounding conductors, 73
    grounding and protection, 76
    underground, 77
  grounding methods, 69
  handling energized equipment and lines, 103
    clearance from live parts, 106
    de-energizing equipment or lines to protect
    illumination levels, 104, 105
  illumination, 79
    attachment plugs and receptacles, 80
    emergency lighting, 80
    fixtures, 80
    receptacles in wet locations, 80
    under normal conditions, 79
  installation and maintenance of electric supply stations and equipment, 78
  installation and maintenance of overhead electric

INDEX 277

[National Electrical Safety Code]
  installation and maintenance of underground electric supply and communication lines, 89
    communication protective requirements, 90
    grounding of circuits and equipment, 90
    induced voltage, 91
    inspection and tests of lines and equipment, 89
    installation and maintenance, 89
    other requirements, 91
  lines and equipment, 92
  operation of electric-supply and communications employees, 102
    operating routines, 99
    protective methods and devices, 93
    supply and communications systems, 92
    supply systems rules for employees, 95-99
    vehicular and pedestrian traffic, 102
  point of connection of grounding conductor, 69
    alternating current, 70
    current in grounding conductor, 72
    messenger wires and guys, 71
  protective arrangements in electric supply stations, 78
    electric equipment, 79
    enclosure of equipment, 78
    rooms and spaces, 79
  purpose and scope, 67, 68, 69
  supply and communication lines, 82, 89
    arrangement of switches, 85
    grounding of circuits, supporting structures, and equipment, 83, 90
    inspection and tests of lines and equipment, 82, 89
    joint use of structures, 88
    other stipulations, 88
    relations between various classes of lines, 85

O

Occupational Safety and Health Act (OSHA), 3, 127, 128
  design safety standards for electrical systems, 129
  definitions, 178-198
  electric utilization systems, 129-131
  general requirements, 131-137
  hazardous locations, 169-171
  special systems, 172-178
  specific purpose equipment and installations, 160-169
  use, 148-160
  wiring design and protection, 137-148
  wiring methods, components, and equipment for general
  general, 129
Ohm's law, 6, 9

P

Perceptible currents, 6, 7

R

Release currents, 8
Resistance at contact surfaces, 9
Resuscitation, 10, 12

S

Safe work practices for hazard control, 21, 22
Safety criteria for design and construction, R&D facilities, 228
Safety criteria for general operation of an R&D facility, 231
Safety criteria for working on electrical equipment in R&D facility, 232
    classes of electrical hazards, 233, 234
[Safety criteria for working on electrical equipment]
    minimizing risk, 232
    minimum personnel and administrative requirements, 235, 236
    modes for working on electrical equipment, 234, 235
    planning the work, 233
Safety rules, general, 23
    electronic equipment, 25
    equipment design, 31
    impulse currents: condensers, 27
    outdoor locations, 30
    power distribution, 24
    power tools, 25
    rf and microwaves, 29
Safety review committee, 201

V

Ventricular fibrillation, 1, 10, 15

Practical Electrical Safety

D. C. Winburn

Los Alamos, New Mexico

MARCEL DEKKER, INC.　　New York and Basel

ISBN 0-8247-7948-7

Copyright © 1988 by MARCEL DEKKER, INC. All Rights Reserved

Neither this book nor any part may be reproduced or transmitted in any form or by any means, electronic or mechanical, including photocopying, microfilming, and recording, or by any information storage and retrieval system, without permission in writing from the publisher.

MARCEL DEKKER, INC.
270 Madison Avenue, New York, New York 10016

Current printing (last digit):
10 9 8 7 6 5 4 3 2 1

PRINTED IN THE UNITED STATES OF AMERICA